A Framework for Community Ecology
Species Pools, Filters and Traits

This book addresses an important problem in ecology: how are communities assembled from species pools? This pressing question underlies a broad array of practical problems in ecology and environmental science, including restoration of damaged landscapes, management of protected areas and protection of threatened species. This book presents a simple, logical structure for ecological assembly and addresses key areas, including species pools, traits, environmental filters and functional groups. It demonstrates the use of two predictive models (CATS and Traitspace) and consists of many wide-ranging examples, including plants in deserts, wetlands and forests, and communities of fish, amphibians, birds, mammals and fungi. Global in scope, this volume ranges from the arid lands of North Africa, to forests in the Himalayas, to the Amazonian floodplains. There is a strong focus on applications, particularly the twin challenges of conserving biodiversity and understanding community responses to climate change.

Paul A. Keddy is an independent researcher who has taught community ecology for more than 30 years. Paul has conducted award-winning research on environmental factors controlling plant communities and their manipulation to enhance native biodiversity. He delights in bringing science alive for his audience. The author of six books winning three scientific prizes, he also co-edited *Ecological Assembly Rules: Perspectives, Advances and Retreats* (Cambridge University Press 1999).

Daniel C. Laughlin is a professor of plant ecology and ecological modelling in the Department of Botany at the University of Wyoming. Daniel's research focuses on developing quantitative approaches to understand and predict how plant species and communities respond to global change. His lab develops trait-based models that translate ecological processes into statistical frameworks to predict how communities assemble along environmental gradients and how species interact at local scales.

A Framework for Community Ecology

Species Pools, Filters and Traits

PAUL A. KEDDY
Independent scholar

DANIEL C. LAUGHLIN
University of Wyoming

CAMBRIDGE
UNIVERSITY PRESS

CAMBRIDGE
UNIVERSITY PRESS

University Printing House, Cambridge CB2 8BS, United Kingdom

One Liberty Plaza, 20th Floor, New York, NY 10006, USA

477 Williamstown Road, Port Melbourne, VIC 3207, Australia

314–321, 3rd Floor, Plot 3, Splendor Forum, Jasola District Centre, New Delhi – 110025, India

103 Penang Road, #05–06/07, Visioncrest Commercial, Singapore 238467

Cambridge University Press is part of the University of Cambridge.

It furthers the University's mission by disseminating knowledge in the pursuit of education, learning, and research at the highest international levels of excellence.

www.cambridge.org
Information on this title: www.cambridge.org/9781316512609
DOI: 10.1017/9781009067881

First published 2022

A catalogue record for this publication is available from the British Library.

ISBN 978-1-316-51260-9 Hardback
ISBN 978-1-009-06831-4 Paperback

Contents

Preface

Before diving into the first chapter, take a moment to enjoy the acrylic painting *Rambling Rio Grande* by Johnathan Harris, reproduced on the cover of this book. This iconic depiction of the arid southwestern USA illustrates many of the themes in this book.

Let's start at the river and work our way up in elevation. The riparian zone at the river's edge is dominated by fast-growing plant species. Even though the surrounding region is high desert, the community of riparian plants thrives in the fluvial sediments that have been deposited near the edge of the Rio Grande. This vegetation consists of wet sedge meadows and cottonwood forests with understories of Coyote Willow (*Salix exigua*), Seepwillow (*Baccharis wrightii*), and Indigo Bush (*Amorpha fruticosa*). Yet, outside the floodplain, soil moisture plummets, and these riparian species are replaced by Pinyon–Juniper woodlands. Pinyon Pine (*Pinus edulis*) and One-seed Juniper (*Juniperus monosperma*) are short trees that tolerate dry conditions because of the physiological properties of their leaves and wood – that is, their "traits." Climbing the elevation gradient further, we enter a narrow band of mixed forest dominated by Ponderosa Pine (*Pinus ponderosa*), Douglas Fir (*Pseudotsuga menziesii*) and Quaking Aspen (*Populus tremuloides*). This taller and denser forest is supported by summer thunderstorms and deeper snowpack. Beyond the tree line in the purple mountains, cushion plants like Moss Campion (*Silene acaulis*) and other alpine species tolerate the wind, the bitter cold winters, and the short growing seasons. Thus, this painting depicts distinct communities that arise along one elevation gradient.

The goal of this book is not merely to describe such patterns, but to lay out a general causal framework that explains how they arise. Both the causal framework for ecological communities, and the ecological models for assembly that we introduce in this book, are, to our mind, similar to *Rambling Rio Grande*, in the way they depict the essential features of ecological landscapes in a simplified yet vivid form.

How are communities assembled from species pools? This pressing question underlies a broad array of practical problems, including restoration of damaged landscapes, management of protected areas and protection of threatened species. In this book we present a logical structure for ecological assembly and address key topic areas, including species pools, traits, environmental filters and functional groups. After laying the foundations, we also explore two models of trait-based community assembly (CATS and Traitspace) to translate concepts into predictions. Our examples are wide-ranging, including fish, amphibians, birds, mammals, fungi and plants from the arid lands of North Africa to forests in the Himalayas and Amazonian floodplains. There is a strong focus upon applications, particularly the twin challenges of conserving biodiversity and adapting to climate change. We emphasize throughout that modern ecology is built on older foundations and remind readers of important lessons from ecologists, including Wallace, Tansley and Rigler.

The authors first met in Flagstaff, Arizona in 2007, when Daniel invited Paul to give a seminar about his research on coastal wetland restoration. At the time, Daniel was completing his doctoral studies on long-term vegetation dynamics in the ponderosa pine forests of northern Arizona. This visit allowed a side trip to the deserts of Arizona to enjoy Saguaro Cacti (*Carnegiea gigantea*) and Phainopeplas (*Phainopepla nitens*). A lot has happened since. Daniel moved to New Zealand to teach and study plant ecology at the University of Waikato. Paul left his position as an endowed professor of coastal restoration in Louisiana, returning to Canada to become an independent scholar. In 2017 Daniel returned to North America to begin a new chapter in Wyoming. Along the way he made an epic trip across the continent and stopped in Ontario for a visit with Paul. This allowed for forest excursions and more conversation. Daniel is now a professor of plant ecology in the Botany Department of the University of Wyoming, where he is conducting research on the ecology and restoration of western US ecosystems in the face of climate change. Paul has continued living in the forest (many years longer than Thoreau, now gaining on St. Francis), where he continues pursuits in ecological synthesis and research in conservation.

The photo below shows us on the front steps of Paul's home, shortly after agreeing that it was time for a new book on the assembly of ecological communities. We thank our lifelong partners, Cathy Keddy and Kara Laughlin, for their unwavering support and laughter throughout this project.

Paul Keddy, Daniel Laughlin

I A General Framework for Community Ecology

The community vector **C** *as a subset of the species pool* **P**. *Landscapes, communities and quadrats. Assembly and response. Foundations from Raunkiaer, Major, Mueller-Dombois and Ellenberg. The elements of community ecology: pools, filters, traits, dispersal, time.*

COMMUNITIES

Every ecological community, be it a coral reef, a eucalyptus forest, a tract of prairie, or an alpine lake, is comprised of a set of species. Those with a keen eye for wild nature often see what seem to be the same sets of species recurring, particularly when similar causal factors such as flooding, fire or drought are present. The study of community ecology, therefore, begins with, and is focused upon, one key descriptor: the community vector, which gives the abundance of a set of species in a selected location (Figure 1.1). Examples of communities include the birds found on a tropical island, the fish found in a river delta, the amphibians found in a vernal pond, the plants found in a hectare of Appalachian forest or the beetles occupying an animal carcass.

The scientific literature of ecology, and the unpublished field notes and data files of natural historians and ecologists, are full of such lists of species. In its simplest form, this vector is just a species list, with occurrences scored as either 0 (absent) or 1 (present). Often, too, field observations include more detailed measures of abundance for each species. Probably the best measure of abundance, overall, is biomass. For some organisms, such as birds or beetles, it may be equally acceptable to use the number of individuals.

$$C \subseteq P$$

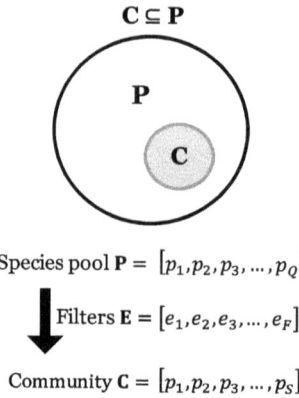

Species pool $\mathbf{P} = [p_1, p_2, p_3, \ldots, p_Q]$

Filters $\mathbf{E} = [e_1, e_2, e_3, \ldots, e_F]$

Community $\mathbf{C} = [p_1, p_2, p_3, \ldots, p_S]$

FIGURE I.I   The vector **C** gives the abundance of each species within a specified sample unit, be it an island, a quadrat or a pitfall trap. It is the central data set in community ecology. Each region of Earth has a pool of species, **P**, each of which might occur in our sample unit, but normally only a subset of this pool, **C**, is found. The subsetting action is driven by the environmental filters, **E**.

The habitat that the vector represents can be any selected area of space. At one extreme, the habitat may consist of a discrete and very physical island, as in *The Theory of Island Biogeography* (MacArthur and Wilson 1967), or units of landscape such as mountaintops or lakes that have many of the properties of islands. At the other extreme, the vector may represent a rather small sample unit of arbitrary size that is delineated within a larger area of continuous habitat. Such smaller sample units may include the familiar quadrat, as well as the contents of a plankton net or pitfall trap, or the enumerated list of species observed on the ground within a single fallen log or high in the canopy in a water-trapping bromeliad. Sometimes, therefore, the vector represents a quite small and narrowly defined location, whereas other times it represents a larger area.

With such a vector in hand, many of the essential aspects of community ecology emerge. We can ask how many species are in the vector, and why it should be so (e.g., Hutchinson 1959, 1961). We can ask what causal factors are responsible for the abundances we observe (e.g., Grime 1979, Keddy 2010). Note that the term *causal factors* can include both abiotic factors (flooding, drought, fire) or biotic factors (mutualism, predation, competition) (HilleRisLambers et al. 2012). We can calculate a measure of diversity (e.g., Pielou 1975, Magurran and McGill 2011). We can also take a set of such vectors, creating a species by sample matrix, and explore patterns among them, taking us into the enormous realm of ordination and classification techniques (e.g., Gauch 1982, Digby and Kempton 1987). We can also sort the species into groups with similar characteristics, thereby simplifying the vector from S species into a shorter list of functional types (e.g., Grime 1977, Cummins and Klug 1979). We could therefore say that community ecology is fundamentally the study of patterns in one or more vectors of species data.

In this book, we do not intend to summarize all the activities that you can carry out with one or more such vectors, as this would require a textbook summarizing all of community ecology (Mittelbach and McGill 2019). Indeed, there is a huge literature on each causal factor that could control a community, with drought and competition being familiar examples. We will focus instead upon two questions that could be considered the central challenges of community ecology. We call them assembly and response rules. Both involve prediction, which ecologists like Rigler and Peters remind us is an essential component of ecology (Rigler 1982, Rigler and Peters 1995). But first let us digress to put these general principles into a practical context.

A GROUNDING IN REALITY

If this discussion all seems somewhat theoretical so far, let us ground it solidly in biological reality. Let us think about these concepts from the practical perspective of how we can wisely manage natural areas such as national parks and ecological reserves. We use this example because protecting a global system of reserves linked by ecological corridors is one of the conservation challenges of modern times and is likely to continue to challenge ecologists in coming generations. Most reserves contain one or more natural communities, each produced by some set of casual factors. You can think about a particular natural area that is important to you, and see if you can find a map of the ecological communities therein. Or, you can enjoy the example in Figure 1.2, Kruger National Park in southern Africa, one of the largest protected areas in Africa (ca. 20,000 km^2).

In Kruger National Park, Gertenbach (1983) divided the park into 35 "landscapes" based upon a combination of physical properties, such as soil type, and biological properties, such as the fauna. Thirty-five may seem like rather a large number of different kinds of communities within a single park, and indeed it is larger than the number of types illustrated in Figure 1.2. But in fact, some ecologists might reasonably observe that these landscape types are too large in scale and too variable in composition to be called communities at all. Here is a summary of Gertenbach's description of the first landscape type.

"Lowveld Sour Bushveld of Pretoriuskop" is a landscape under-lain by granite and gneiss, producing acid soils. (Hence, the term "sour." We shall see another granite and gneiss terrain on a different continent in Chapter 2.) You can see the "sour bushveld" landscape shown as a stippled area on the lower portion of the map (Figure 1.2). The vegetation consists of open tree savanna with low shrubs. There is also a rich herbaceous flora. The relative abundance of woody plants and herbaceous species changes with soil moisture. The animals include Reedbuck, Kudu, Northern White Rhinoceros and Sable Antelope.[1] There are 530 km^2 of this landscape type in the national park.

Within each such landscape, we could indeed look for more narrowly defined sets of habitats with characteristic sets of species. In fact, this raises a difficult question for community ecology: just

---

[1] For groups with well-established common names, like birds, we use them alone. Where we judged there might be uncertainty (e.g., in many groups of plants or insects), we have included the Latin name. There is not a perfect solution, particularly given multiple human languages, and the large numbers of species in certain taxonomic groups. The same approach has been used in the index. We also capitalize common names, against the trend in convention, because while there are many warblers that are yellow and frogs that are green, there is only one Yellow Warbler and one Green Frog.

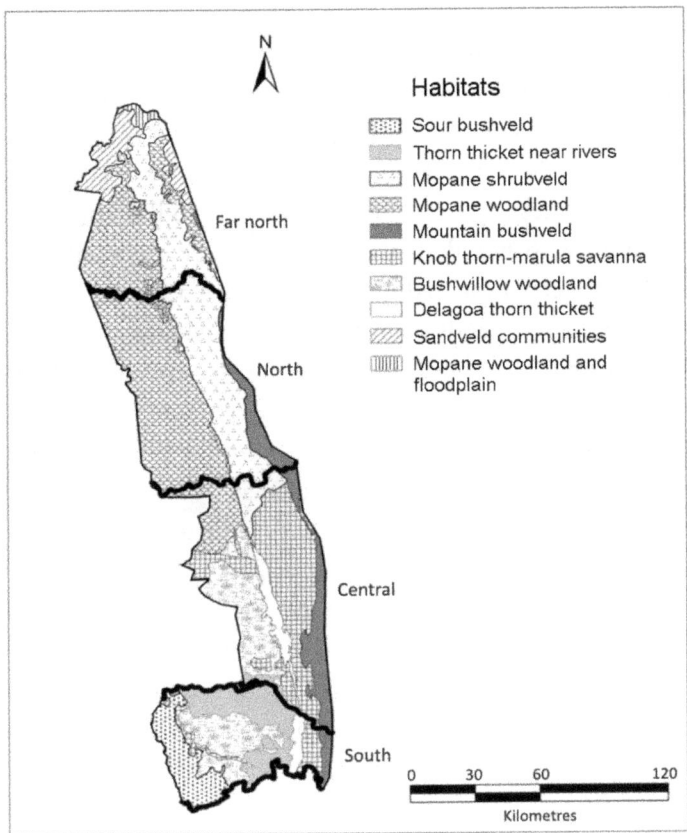

FIGURE I.2 The major "habitats" of Kruger National Park in southern Africa. The raw data were the 35 different "landscapes" documented by Gertenbach (1983). Chirima et al. (2012) simplified these "landscapes" into just eight major "habitats," but did not include the northern 40 km of the park. Our map is expanded from theirs to show the entire park, which necessitated adding two more habitats based on Gertenbach's map (adapted from Chirima et al. (2012) and Gertenbach (1983) by Cathy Keddy).

when do we stop the process of subdividing the landscape into smaller scale units, and how small, and how uniform in composition, do these smaller subdivisions have to be for us to call them a "community?" We are going to deliberately skate around this topic. At first it might seem like a reasonable question to ask. It is also quite possibly the kind of question that causes confusion rather than providing clarity. We might suggest that some of the problems in the scientific development of community ecology have arisen precisely because the wrong questions have been asked. Recall the paradox of Zeno's tortoise, which turns out to not be very useful if you really want to know how long it takes a tortoise to walk from A to B. Zeno was a Greek philosopher who noted that the tortoise will never actually reach point B if you continue to divide the distance between the two points in half. That is, one ends up dividing a finite distance into an infinite number of small distances, which may be a useful analogy to some of the activity in community ecology, where even grant size may vary inversely with distance covered. It also implies that it is safe to jump off tall buildings, since, like Zeno's tortoise, you will never quite reach point B, in this case, the pavement. Returning to Kruger National Park (which, by the way, has several tortoise species, including the magnificent leopard tortoise), Gertenbach describes how there is indeed much more fine-scale data on Kruger National Park, including some 1,500 vegetation quadrats collected using the Braun–Blanquet technique. He concludes that the decryption and subdivision into communities at such a fine scale has been of doubtful use to park management at all:

> The geographical distribution and size of these plant communities have shown such a complex pattern as to render it impossible to indicate individual plant communities or associations on a map of reasonable scale. Despite intensive research regarding the ecology of these plant communities, no practical management program based upon these communities has been forthcoming. The intensity of these surveys has thus apparently surpassed the practical application of the results.
>
> (p. 9)

In our view, for science in general – and community ecology in particular – we think it is necessary to consider practical applications. This is true for two rather different reasons. First, of course, we want to be able to protect wild species and we need practical classifications of communities and habitats to do so. That is partly what motivated both of us to become ecologists in the first place. But there is a more fundamental reason. If we do not demand practical applications from our science, it is easy for that body of science to drift off into intellectual obscurity (Rigler 1982, Peters 1992). You may recall that medieval scholastics were notorious for being able to devote time to arguing over how many angels can dance on the head of a pin. Galileo, in contrast, thought about trajectories of cannonballs rather than angels, which has led us to some remarkable celestial truths. A critic might complain that too much community ecology comes close to being scholastic (e.g., Keddy 1987). By thinking about real examples, and demanding practical consequences, we force ourselves into a different sort of rigour. Since collecting data costs professional time and money, and since we live in a world with limited amounts of both, and a world with multiple competing demands, we really want to avoid creating situations where the "intensity of our surveys surpasses the practical application of the results" (Gertenbach 1983).

Here is another example of practical issues in designating communities. If you really need to have a practical classification system, you do have to address the issue of scale. In Ontario, for example, there is an ecological land classification hierarchy: Ontario is divided into ecozones at the provincial scale. These are then further subdivided into ecoregions and ecodistricts (if you flip ahead, you can see a map of ecoregions and ecodistricts in Figure 3.2). These in turn are subdivided further into ecosections, ecosites and ecoelements. Note that these authors do not use the word community at all! Many ecologists would likely apply the word community to either of the two lowest levels. Either level could be reasonably described by a vector of species abundances. The authors themselves (Crins et al. 2009) say that ecoelements are more or less equivalent to vegetation types, while ecosites are "Fine-scale landscape areas defined by recurring patterns of ecoelements."

The important point is this: any particular landscape, or protected area, will have such a nested hierarchy. The community lies toward the lower end of this continuum, whatever term a particular classification system uses. And, as Gertenbach observes, if we move too far down the scale, we reach a level of so much detail that practical application may be obscured.

COMPOSITION AND CAUSAL FACTORS CAN BOTH BE MAPPED

We can consider any protected area as having a set of communities (if we are thinking about it from the point of view of the vector **C**) or we can talk about a protected area as having a set of habitats (if we are thinking about it from the point of view of the causal factors that produce the vector **C**). On one hand, this is obvious. We can call an area of our landscape a floodplain community because we have data from the location showing the presence of flood-tolerant trees, or we can call this area a floodplain habitat because we have data showing that the location is under water for several months each spring. Neither is wrong, but each depends upon a different kind of observational data. On the other hand, we have to try to use words carefully. Mapping observed communities is quite different from mapping causal factors. In this book, we shall try to use the word *community* mostly to refer to the observed species composition (the vector **C**), while we use the word *habitat* to refer to a set of causal factors. In general, any location has a set of causal factors that can be ranked in order of their influence upon the community. In wetlands, for example, flooding is the key causal factor, but other factors such as fertility, natural disturbance, grazing, burial and salinity act to modify the effects of flooding and thereby determine the particular kind of wetland that arises (Keddy 2010). Causal factors can also be biological. In wetlands, again, grazing by animals can be an important causal factor controlling plant composition. And, for many animal species, biotic factors can also be causal. As we shall see in Chapter 2, both amphibian and fish communities may both be strongly affected by predatory fish, in which case predation becomes a causal factor. Similarly, the abundance and structure of woody plants is a well-known causal factor in bird communities: there are clear differences between birds of grasslands and birds of forests, even at quite local scales. We will mostly use the word *filter* to refer to a single causal factor, and, in the next chapter, we will describe what filters are and how they work in landscapes and communities.

THE LANGUAGE OF SAMPLES AND SAMPLE UNITS

Most areas of landscape, particularly those in protected areas like Kruger National Park, have maps showing the patterns that occur, be they "landscapes," "communities," "forest types," "habitats" or "vegetation types." Sometimes reserve managers have an exhaustive and comprehensive list of all the species in each of these categories. More often, what they have is a set of sample units from each community (or habitat). That is, the communities that you may see on a map of the reserve are documented by only a sample of some number of sample units. This is usually a simple biological and financial reality: there is only so much money and time, so any particular type of habitat is described by collecting data from a set of quadrats, a set of lists from bird counts, a set of data from fish traps and so on. For the purposes of communication, note that people are often careless about the use of the word "sample." In this book we shall use the word sample to mean a set of observations. A single observation, like a bird list or the contents of a pitfall trap, is a sample unit. The point is that in even relatively small tracts of land, our knowledge of the species composition is derived from a sample, or, that is, a set of sample units. In the case of Kruger National Park, there were 1,500 quadrats (i.e., sample units) documenting vegetation.

Hence, when you visit a particular location within such a protected area, you are viewing one sample unit. There is a larger vector that describes all the sample units combined, that is the community of organisms documented from all the samples collected in that habitat. If we put in enough effort, and collect enough samples, we should end up with an exhaustive and complete list of our species pool. We shall discuss this process further, and give examples, in Chapter 3 on species pools.

MORE GENERAL PRINCIPLES: ASSEMBLY AND RESPONSE

Now let us move back to general principles for community ecology. We said above that there are two fundamental questions. Our first challenge is to predict the vector **C** for a given location based upon the environmental filters present at that location, or, in other words, the measurable properties of the habitat. Examples of this challenge include: which birds will be found on this year's Christmas bird count? Which tree species will be found in a tract of dipterocarp forest in Borneo? Which species of frogs will you find in a vernal pool in prairie parkland? In most cases, we may know the potential list of species, say **P**, that might be present, from standard documents such as floras and wildlife guides, which provide what we shall call the species pool. The challenge, then, is to predict which set of species **C** (and their abundances) will actually be found at a specific location (Figure 1.1). That is our primary goal in this book.

There is also a second kind of question (Figure 1.3). Assume that we have a specified location in space, at time t, with a list of abundance of each species present ($\mathbf{C}_t$). Now, assume that a filter is

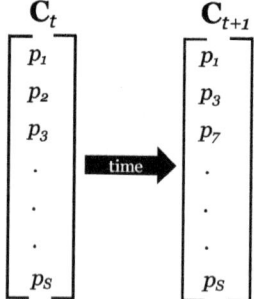

FIGURE 1.3 In some cases we know the current vector of species abundances $\mathbf{C}_t$, but the challenge is to know what the vector will become after some specific change in one or more environmental filters. Our challenge is to predict $\mathbf{C}_{t+1}$.

going to be changed: for example, temperature, rainfall, grazing intensity or predation. How will **C** respond to that changed filter? What will C_{t+1} look like? This style of question underlies many current environmental questions, including how forest composition might shift with changing climate, or how fish composition in estuaries might shift with changing fishing pressure, or how forests will respond to logging. Indeed, we could say that this question is the scientific underpinning of environmental impact assessments overall. A wide array of applied questions about the future therefore share a common structure. How can we use knowledge of pools and traits to predict community responses to changing filters? We call this class of problems response rules, as distinct from assembly rules. Although the focus of this book is on assembly rules, our intention is also to consider to what degree response rules may benefit from a similar kind of thinking.

This second challenge, response rules, may appear more difficult, because we are being asked about the response of the set of species to a change in environmental conditions. However, this question also includes one added piece of useful information: the existing state of the system. We are not being asked to predict *de novo* from a very large number of possible species; we have prior information on the raw material that is already present in the location. So, the extra demands of prediction may be balanced by the prior information on species that are already present, and their traits.

We view the construction of such assembly and response rules as the central, and defining, issue of community ecology. In this book, we will largely focus on assembly rules. We will pursue one particular approach to this problem. It involves knowledge of species pools, functional traits and environmental filters. This is, perhaps, not the only framework for creating assembly and response rules, but it is the one which we believe holds the most promise for unifying community ecology and for providing the vital ability to predict the composition of ecological communities altogether (Weiher and Keddy 1999b, McGill et al. 2006).

The scope of the problem is demonstrated by the complexity of the challenge. Earth has over one million species of organisms. If we focus upon plants alone, since they comprise the majority of Earth's biomass, and provide energy and habitat for nearly all other species, there are still approximately 350,000 species (The Plant List 2013). Anyone who has compared a desert to a wetland, or a lakeshore to a mountaintop, will appreciate that these 350,000 plant species do not occur in random mixtures. Certain groups of species tend to occur as repeating patterns. Indeed, we recognize deserts partly because of dominance of succulents, while we recognize wetlands because of the presence of flood-tolerant plants. Similarly, if you take a canoe trip down a river, or hike along a chain of mountains, you see that certain combinations of species occur again and again.

So, at one scale, it is obvious that certain environmental constraints do indeed select certain vectors of species composition from the global pool of 350,000 species. That, too, is why we can recognize biomes and ecoregions. It is also true, however, that demonstrating patterns in species composition is a challenge at smaller scales, and there is an enormous scientific literature on the challenge of finding recurring patterns in species composition. One might indeed argue that the vast array of papers on patterns in ecological communities has become a distraction of sorts, since there really is no end to asking whether there are patterns in species composition of particular taxa in particular habitats at particular scales. More importantly, the search for such patterns can become an end in its own right, subtly distorting the search for predictive rules based upon cause and effect. Without caution, the organizing theme in ecology can degrade to a simplistic question: is there a pattern? Even if there is a pattern, such information puts one only marginally closer to knowing why, or more importantly, to being able to predict vectors from known causal factors. We will explore this problem more in the next section.

THE STUDY OF PATTERN IS NOT THE STUDY OF COMMUNITY ASSEMBLY

Since there seems to be considerable confusion on the topic of pattern in ecological communities, let us look at just how much effort has gone into detecting pattern in plant communities over a period of more than 50 years. Although we begin with the example of plants, let us make it clear we are after the general implications for ecology as a whole. Zoologists have similar problems.

Here is an introduction to pattern in plant communities by Goldsmith and Harrison (1976, pp. 113–114):

> Non-randomness in vegetation is often referred to as pattern. It may take the form of an aggregation of individuals known as a contagion, or an even distribution known as a regularity. The former is more common than either the latter or randomly distributed individuals. The departure from randomness interests the ecologist because it is a way of characterizing the vegetation or a particular species. Also because it must have a cause and it provides an opportunity to identify the factors that control the distribution of a species.

The most common kind of pattern study (and there were many of them) took data from a large number of quadrats, and asked whether certain species were associated with one another, using simple tests such as chi-squared tests (Cole 1949, Greig-Smith 1952, 1957, Kershaw 1973). Usually certain species were found to be associated, and traditionally such papers ended with speculation about what ecological factors might be involved in producing the non-randomness. Dale (1999) has provided an overview of this field.

Another group of studies used data collected on zonation patterns. Pielou (1975) developed a statistical test for asking whether zonation patterns deviated from null models. Since collecting data on zonation patterns requires considerably more work, there are fewer examples (e.g., Pielou and Routledge 1976, Keddy 1983, Shipley and Keddy 1987, Hoagland and Collins 1997). Again, the results were clear in one way: there is statistically significant deviation from null models. But just as with quadrat data, there are multiple hypotheses for what might cause the non-randomness (e.g., Shipley and Keddy 1987). Keddy (2017, pp. 441–448) has provided an overview of this approach.

Then, of course, there is a vast scientific literature on patterns that can be shown using multivariate methods, with a wide array of techniques including reciprocal averaging and detrended correspondence analysis, to name a few (e.g., Orloci 1978, Gauch 1982, Digby and Kempton 1987). These can provide elaborate displays of multidimensional patterns in vegetation (or in other kinds of ecological data), although, unlike simple chi-squared tests, usually do not provide an explicit test for non-randomness. Usually, the non-randomness is assumed, and the techniques are then applied to extract the assumed patterns. Multivariate tools can be useful for describing large data sets, but too often they became an end in their own right, and a subtle justification for collecting large data sets. Paul once overheard a fellow scientist say his career objective was to provide an ordination description of every vegetation type in his province.

The point is that the search for pattern can become an end in itself, and in most of the papers one reads, questions about causal factors are relegated to the end of the paper for speculation in the discussion. Anyone the least familiar with the ecological literature will immediately recognize how many papers fall into this category. After all this work over the past 50 years, we can conclude with certainty that certain species, in some locations, at some scales, show non-random patterns. The problem is that we cannot predict with any confidence which species, which locations or which scales will have which kind of pattern. The plant ecology story is intended to be a cautionary tale, because it would be possible to go through the same process with any other group of organisms, be it birds on islands or fungi on fallen logs or parasites in animal guts, and with probably the same outcome: patterns in some species, at some locations, at some scales.

For progress in ecology, we therefore have to make a concerted effort to move away from simply looking for pattern, and instead start looking for predictive ability (Keddy 1987, Peters 1992, Weiher and Keddy 1995). That is our goal in this book. Our first task is to illustrate some general steps that provide a logical framework. Otherwise, it might be easy to throw up our hands and give up on prediction, and simply go out and collect more observations. Or, worse still, to host yet another conference on emerging paradigms in community ecology. One is nearly as pointless as the other. Meanwhile, while we entertain ourselves as scholars, we face the prospect that perhaps a quarter of the world's biota could disappear this century. One might argue that not only is the capacity to predict both the community vector **C** and changes in **C** with time one of the most intellectually challenging parts of community ecology, it is simultaneously one of the most significant moral challenges for protecting Earth's biota.

The design of a global system of protected areas might be seen as one of the most demanding applications of community ecology. Consider the typical questions in the design of a protected areas network. How many species occur in the landscape? Which areas are most important for protecting these species? What are the vectors of **C** for different habitats and sample areas? What are the key factors that enhance **C**? What are the key factors causing the vector **C** to decline in length? How can we restore degraded areas to increase the length of **C**? All of these fundamental questions involve recurring questions about species pools (**P**), communities (**C**), and environmental filters (**E**).

Now let us turn to some historical context for this inquiry.

FOUNDATIONS OF COMMUNITY ECOLOGY LAID BY RAUNKIAER

Consider the work of Raunkiaer on the distribution of plants globally. This work is now over a century old, and yet has several important themes that will recur in this book. First, there is the problem of a very large number of species – that is, a large pool. Second, it is possible to identify functional traits that these species possess. Third, it is possible to sort species into groups based upon these traits. Fourth, there is the challenge of finding patterns. Fifth, and most important, there is the challenge of relating species to environments based upon causal factors. Raunkiaer's work illustrates all of these themes. Raunkiaer is chiefly remembered for his functional groups, but here we wish to draw attention to the other important aspects of his work, particularly the insight of linking functional traits to environmental filters.

At the time he was working in the early twentieth century, ecologists had become aware that there were vast numbers of species, and that the world was much more complicated than earlier generations of European scientists had understood. To put his work into context, recall that the epic voyage of the *Beagle* had begun in 1831, while Captain Cook's voyage to Australia had occurred in just 1770. Clearly the genera recognized by Linnaeus were insufficient for cataloguing the diversity of plants and animals. What generalizations could be drawn? Raunkiaer proposed that the world's plants could be divided into ten life forms. These were based upon the location of the meristems on the plants (you can have a quick look at them in Figure 6.1). Phanerophytes, for example, are woody plants with buds (meristems) borne above the surface of the ground. He then collected data to show that different environments had different kinds of life forms. He also developed a null model, the "normal spectrum," long before there were statistical models and computers for the task, by selecting plants from a global list of plants called the *Index Kewensis*. He showed that the life forms of plants in his data were non-random (Table 1.1). For example, the top row in the table shows that Franz Joseph Land, an archipelago in the Arctic Ocean north of Russia, had just three of 10 life forms (Chamaephytes, Hemicryptophytes and Geophytes). Overall, Raunkiaer concluded that this variation was a consequence of climate, with cold and drought in particular controlling which kinds of life forms could be found.

Because this work is a century old, it is easy to assume the ideas are out of date, particularly if you are one of those unfortunate students whose professor told you not to read work more than five years old. Yet,

Table 1.1 *The distribution of ten plant life forms in seven locations with different climates (Raunkiaer 1908). The point of the table is to show non-random patterns in an important life history trait, and a null model (the bottom row, the "normal spectrum"). Selected life forms are illustrated in Figure 6.1. More examples can be found in Keddy (2017).*

Region		No. species	Percentage distribution of species among life forms									
			S	E	MM	M	N	Ch	H	G	HH	Th
Franz Josef Land, Russia	82° N, 55° E	25	–	–	–	–	–	32	60	8	–	–
Iceland	65° N, 19° W	329	–	–	–	–	2	13	54	10	10	11
Sitka, Alaska	57° N, 9° E	222	–	–	3	3	5	7	60	10	7	5
Death Valley, California	36° N, 117° W	294	3	–	–	2	21	7	18	2	5	42
Ghardaïa, Algeria	32° N, 4° E	300	0.3	–	–	–	3	16	20	3	–	58
Aden, Yemen	13° N, 45° E	176	1	–	–	7	26	27	19	3	–	17
Seychelles	5° S, 56° E	258	1	3	10	23	24	6	12	3	2	16
Normal spectrum		1,000	1	3	6	17	20	9	27	3	1	13

Abbreviations: S, stem succulent; E, epiphyte; MM, Megaphanerophyte; M, Mesophanerophyte; N, Nanophanerophyte; Ch, Chamaephyte; H, Hemicryptophyte; G, Geophyte; HH, Helophyte and Hydrophyte; Th, Therophyte.

Raunkiaer had shown that plants occur in non-random groups well before all the literature we just reviewed on pattern in ecological communities. He also identified a key causal factor, climate. And he identified an essential functional trait that still remains useful in modern ecology, the position of the meristem. The work still matters. Every time that you see a vast marsh, note that one characteristic is shared by all the dominant marsh species: persistence from buried, not aerial, meristems. Every time you see a tract of forest, note that one characteristic is shared by all those dominant forest species: persistence from aerial meristems. More than 100 years have passed since Raunkiaer's work, and we might therefore challenge ourselves to find similar relationships between traits and filters in other groups of organisms and at other scales.

A GENERAL FRAMEWORK

We will now propose a general framework that builds on Raunkiaer and incorporates decades of thinking about species pools, filters and traits. We are going to borrow key elements of frameworks presented by Jenny (1941) and Major (1951), who focused, on soil pedogenesis and plant communities, respectively. After some rearranging of terms, their general statement reads:

$$\text{plant community} = f(o, \ c, \ p, \ r, \ t) \tag{1.1}$$

where o = organisms, c = climate, p = parent soil material, r = relief or topography, and t = time.

This framework suffers from being inordinately focused upon plants, but it is a good starting point, since plants comprise as much as 90 percent of all the biomass on Earth. (We will revisit this particular statement in Chapter 4, particularly in Figure 4.1.) A similar presentation of causal factors can be found in a book chapter entitled "Casual-analytical inquiries into the origin of plant communities," written by Mueller-Dombois and Ellenberg (1974). The chapter begins with the observation that studies of patterns in community composition do not tell us much about mechanisms. Muller-Dombois and Ellenberg begin with a simple but general observation that applies across all taxa: knowledge about mechanisms that underlie patterns in communities can "only be obtained from measurements and experimentation." It is ironic, and instructive too, that this chapter is 12th in a thick book that spends a great deal of space (11 chapters, 334 pages) talking about patterns in plant communities. This is, of course, part of the problem addressed earlier in this chapter: we have inherited a scientific discipline in which too many people are quite satisfied to focus their attention on description.

Building upon such ideas, let us formalize the various factors that influence the community vector **C**. We begin with the species pool, **P**. We will list them first, and then consider each in turn.

P: the species pool, a vector of Q species, $\mathbf{P} = [p_1, p_2, p_3, \ldots, p_Q]$

C: the community, a subset of **P**, a vector of S species, $\mathbf{C} = [p_1, p_3, \ldots, p_S]$,

E: environmental filters, a vector of F causal factors, $\mathbf{E} = [e_1, e_2, e_3, \ldots, e_F]$

T: functional traits, a vector of K traits, $\mathbf{T} = [t_1, t_2, t_3, \ldots, t_K]$

d: dispersal

t: time

The first three are included in Figure 1.1. We shall add the others into the story later.

THE BASIC ELEMENTS OF COMMUNITY ECOLOGY

Now let us turn to the list presented above, and briefly consider each one by way of introduction to their potential application to assembly rules. Most of the rest of the book will be revisiting these six elements in more detail. What do we mean by **P**, **C**, **E**, **T**, d and t, and how does each apply to the problem of community assembly?

P: The Species Pool

The species pool **P** is determined by the location of the community, and its evolutionary history. It is the raw material from which a community is assembled. **P** varies with location: a fragment of the old continent of Gondwana off the coast of Africa, like the island of Socotra, will have a very different pool from a piece of continental Laurasia in eastern North America. There are evolutionary and geological reasons that pools differ with location (Wallace 1876, Takhtajan 1986, Rosenzweig 1995). To keep it simple, for the moment, just think of your favourite local field guide to the fauna or flora as a good first approximation of the species pool. When you have a bird to identify, your local guide gives you the pool to choose from. We devote a later chapter to the nuances of defining a species pool.

C: The Community

The list of species found in any particular community **C** is the set of species that occur there at time t. It is the list of species, always a smaller subset of **P**, that have arrived, survived and persisted until the time of sampling (Figure 1.1). **C** may be the most common element of assembly rules that ecologists measure. We likely have notebooks full of species lists from habitats we have visited. Indeed, if we think small enough and consider the ubiquitous quadrat as a community, then we have an enormous number of estimates of **C**. We use the word estimate as a humbling reminder that it is always easy to miss species, particularly when they are young or transient, so our list of species is always likely an underestimate of the true **C** in that location. But this is a minor point, since, when you use biomass as a measure of importance, the unseen species are usually small or visually insignificant, and so are a minor component of biomass in any case, at least in sessile communities. We are aware, of course, that in general, big fierce animals are also rare (Colinvaux 1978), and that in any sample, a few species predominate in numbers and mass (Preston 1962). Overall, we have made a good deal of progress in surveying Earth's biota since the days of Captain Cook. We now have huge data sets giving us **C** for a wide array of habitats from quadrats to lakes to tropical islands to national parks.

So, **P** is a list of all possible species that might occur, and **C** is the list of those that actually occur now. How do we get from **P** to **C**? It requires us to know something about the fundamental biology of the species themselves, collectively, and the environmental filters that are acting.

E: Environmental Filters

Every habitat has a hierarchy of environmental filters, which many people also call environmental factors, that control composition. We use the word hierarchy explicitly to emphasize that in any location, some factors are important, and some are unimportant. What are the factors, and how important are they? One would be surprised at just how scarce such information is. Indeed, one might argue that one reason that communities are hard to predict is that too many people study trivial rather than important factors. Filters can be physical or biological in origin. Let us look briefly at three physical filters (climate, moisture and oxygen) and then some biological filters (predation, herbivory and competition). We will explore the topic of filters in more detail in Chapter 2. Here we are just laying down some basic concepts.

Many early ecologists, including the aforementioned Raunkiaer, Jenny and Major, drew attention to climate. Certainly, in our (northern) part of the world, one of the most important factors has to be the low temperature experienced during each winter, usually about −20 °C. This filter is so efficient that it simply eliminates from consideration all woody plants without traits to tolerate cold. Thus many of the tree species that occur in North America (as listed and mapped in the classic book *Silvics of Forest Trees of the United States* (Fowells 1965)) simply cannot survive in our parts of North America. The pool for northern species is much smaller. Cold weather partly determines the species pool at higher latitudes and altitudes (de Bello et al. 2013). For example, when Daniel teaches ecology in the Rocky Mountains, he only has to describe a handful of species to capture the majority of trees that occur in the mountains. His brother Andrew, on the other hand, teaches ecology in the warmer, more southerly mountains of Appalachia. This area contains many dozens of tree species. The irony is that even though Andrew is an ornithologist, he spends more time teaching tree identification than Daniel, his botanist brother, all because of the strength of the climatic filter that acts in their respective regions. When we consider the species likely to establish in habitats in wetlands around the Great Lakes in northern North America, we simply do not have to concern ourselves with Baldcypress (*Taxodium distichum*), nor Sawgrass (*Cladium jamaicense*), nor with American Alligators (*Alligator mississippiensis*) nor Wood Storks (*Mycteria americana*). So, a first generalization would be that one set of environmental filters is usually so powerful and all-pervasive that the filters act at the scale of the pool. These factors create a regional pool. We will return to this topic in Chapter 3.

A second set of environmental factors, or filters, then act upon species that do occur in the regional pool of species. Since all organisms need water, soil moisture is likely a key factor to consider for trees. Another reason that soil moisture matters so much is the reproductive cycle of plants: all must survive the germination stage. Even huge desert plants like the Saguaro Cactus, which have a certain iconic status for the ability to tolerate drought, must none-the-less have water for seedling establishment. Indeed, they may be dependent upon the occasional wet years that occur only a few times each century (Turner et al. 1966).

Turning to another group of organisms, and a different habitat, aquatic animals are often controlled by the supply of dissolved oxygen (Lowe-McConnell 1975, Junk et al. 1997). Some aquatic habitats are structured by the lowest levels of oxygen that occur. Low oxygen levels may occur during the winter, when lakes are shut off from the atmosphere by ice. Low oxygen levels may also occur in warm weather, since warm water dissolves less oxygen, and warm aquatic conditions often accelerate decay, in which case oxygen is consumed by microorganisms associated with decay, creating anoxic conditions. In the extreme case, freshwater and marine coastal zones can become dead zones, where oxygen is so depleted that fish are killed (e.g., Vallentyne 1974, Turner and Rabelais 2003).

Moisture and oxygen are examples of physical filters. As noted above, there are also filters that are biological in origin. In Chapter 2 we will look at several of them in more detail. At this point we can note that predation is a filter in both amphibian and fish communities (e.g., Tonn and Magnuson 1982, Wilbur 1984). Grazing is also a key factor in many plant communities (e.g., Harper 1977, Keddy 2017). And, then, of course, there is competition (e.g., Jackson 1981, Simberloff 1984). The relative importance of these biological filters remains a matter of intense discussion in community ecology (e.g., Oksanen et al. 1981, 1997, HilleRisLambers et al. 2012, Kraft et al. 2015). In general, however, interspecific competition is assumed to play a particularly important role in community assembly by limiting the number of species that share the same set of traits and selecting for species that differ in traits (Pianka 1981, 1983). The observation that similar species are likely to experience the most intense competition goes all the way back to Darwin, and has led to an enormous number of studies addressing the coexistence of similar species (e.g., Hutchinson 1959, 1961), a familiar example being MacArthur's classic work on wood warblers (MacArthur 1958).

If we consider the direct effects of filters, setting aside competition, we might expect, in a simplistic way, that filters will usually produce a community with an underdispersion of traits. That is, filters will force convergence in traits. Continuing with this perhaps simplistic view, competition is different because the general impact of competition is thought to work in the other direction – to create a community with an overdispersion of traits. If these forces balance one another, the result may appear random. Figure 1.4 gives an overview of how communities might be assembled with regard to filters acting on traits modified by interspecific competition. A good many ecological concepts, particularly the concepts of limiting similarity and resource partitioning, fit neatly into this worldview. Whether this view truly reflects the actual assembly of ecological communities is open to debate (de Bello et al. 2009, Mayfield and Levine 2010,

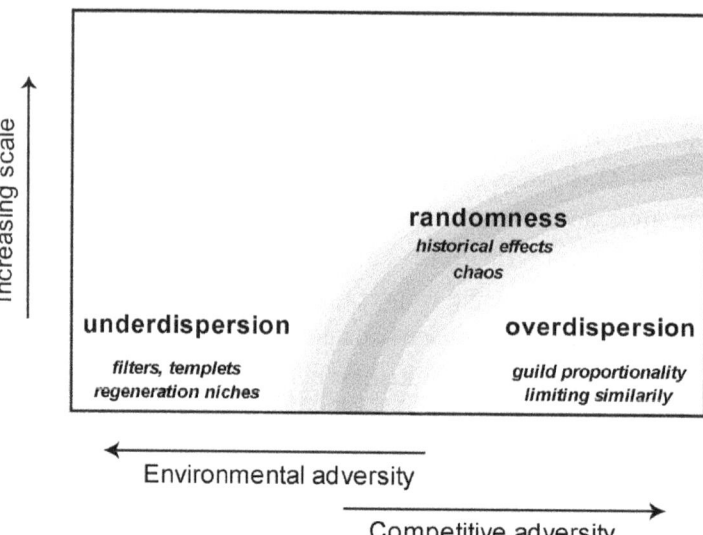

FIGURE 1.4 Most filters select a set of species with similar traits, creating trait underdispersion (left), while competition may select for species having different traits, creating trait overdispersion (from Weiher and Keddy 1995). Overdispersion appears more common at greater spatial scales. This perspective about trait patterns is still common, yet it is an oversimplification of the processes that generate patterns in ecological communities (Mayfield and Levine 2010, Götzenberger et al. 2012, Spasojevic and Suding 2012, Münkemüller et al. 2020).

Götzenberger et al. 2012, Spasojevic and Suding 2012, Kraft et al. 2015, Münkemüller et al. 2020). Let us give two counterexamples.

Let us repeat an essential point about filters. Competition is thought to stand out from all the other biological filters in one important way. While predation or grazing may act like physical filters by selecting for a certain subset of species and traits, competition may be expected to have the exact opposite effect from environmental filtering, by selecting for species that *differ* in traits in order to partition resources (MacArthur 1958, Pianka 1981, 1983). In this context, there is a limit to how similar coexisting species can be. This somewhat traditional view of community assembly is widespread, and shown in Figure 1.4. Warblers can partition space on trees, and finches can partition seed resources based on bill size. Given that animals can partition resources, this view may be more directly applicable to animal communities.

Now let us introduce our first counterexample: tall plants in general, but particularly trees. There are some 60,000 species of trees in the world (Beech et al. 2017). In some habitats, competition may cause a less appreciated and possibly contradictory force to that expected from resource partitioning. Competition may drive coexisting species to *converge* on certain traits, particularly in communities where resources are exploited rather than partitioned, and in communities where space itself is a resource (Yodzis 1978, Keddy 2001). Consider, for example, plant communities in general, and forests in particular. Here, tall species are often better at exploiting light than are short species. Traits for height have their origins deep within the functional ecology and phylogenetic history of land plants. The evolution of height has been a consistent trend, involving several key traits: apical dominance replaces dichotomous branching, and xylem elements replaces tracheids, while cambium with secondary growth allows for thick woody trunks that support foliage for centuries (e.g., Foster and Gifford 1974, Niklas et al. 1983, Kenrick and Crane 1997). These traits improve access to light, and simultaneously suppress smaller neighbours. In this case, however, competition is driving a convergence in life forms, toward tall species with secondary growth.

More generally, there may be some communities that are structured as competitive hierarchies (a topic we will consider further in Chapter 8). In these communities only a few species tend to dominate the community (Keddy and Shipley 1989, Keddy 2001, chapter 5). The dominant species share traits that allow them to suppress most other species. In such situations, coexistence, particularly coexistence of species that are relatively less common, depends upon recurring natural disturbances, such as fire or storms (Grime 1973a, 1979, Connell 1978, Huston 1979). These disturbances create non-equilibrium communities in which many species coexist within a matrix created by a few dominant species. In such communities, the dominant species frequently share similar traits. We have mentioned plants, where height is an obvious trait, but other examples of competitive hierarchies can be found in animal communities, including corals (Yodzis 1978) and even fruit flies (Gilpin et al. 1986).

Now consider our second counterexample to Figure 1.4: aquatic plants. Flooding is a strong environmental filter. We know this because only a small fraction of the world's plant species can tolerate flooding. So it would be reasonable to expect strong trait convergence in response to this filter. At first, it appears that this pattern is commonly observed: most aquatic plants share a small set of traits, including aerenchyma, floating leaves, turions and rhizomes (Sculthorpe 1967, Hutchinson 1975). Yet, when we look more closely at actual communities of aquatic plants, we also find some trait divergence. Instead of one life form, one frequently observes multiple growth forms. Yes, there are floating-leaved plants with deeply buried rhizomes, but also emergent plants, submerged rosette plants, and free-floating plants. The number of functional types of wetland plants that occupy flooded sites can include as many as 26 types (Hutchinson 1975, table 8)! Alright, you might say, perhaps flooded habitats are unique in some way. Yet, if we turn to the opposite extreme, arid lands, where drought is a powerful filter, we find a similar perplexing situation. Yes, drought produces convergence in plant traits, but then when we look at desert plant communities, we find that desert communities can contain a dozen functional types, each representing a unique solution to establishing and surviving in water-limited environments (more on this in Chapter 2).

The point is that the view presented in Figure 1.4 may make general sense, but it is not sufficient. Competition can apparently produce trait convergence (e.g., trees) in communities, while evolution can apparently produce more than one evolutionary solution (set of traits) in response to even strong filters like drought and flooding.

Returning to the general topic of how filters affect communities, we can conclude that each habitat may have a long list of environmental filters that affect assembly. Some are easy to measure and some less so. Some are more important than others. These filters may also vary with space and time. The challenge for community assembly is to find the shortest list of **E** key environmental filters that are currently eliminating most species from **P** and thereby having the greatest impact upon **C**. This is the problem we focus on in Chapters 5 and 7, where we describe the quantitative tools that are available to test these ideas using real communities. And overall, this remains a principal goal of the book: to explore the methods for assembling communities from data on pools and environments. At the same time, the counterexamples just presented will continue to hover on the horizon, suggesting that the larger picture still has some open questions in ecology and evolution. Thus, Figure 1.4 is a useful starting point, but trees and aquatic plants (and arid land plants and corals) show that this figure is incomplete.

T: The Traits of the Species

Every species has a set of traits that allow it to survive in a particular environment. We can compile these into a matrix of Q species by K traits, which we call a trait matrix. Often, we have rather little information to put in this matrix. Consider the importance of flooding, an environmental factor that creates distinctive communities known as wetlands. Which species might be most tolerant of flooding? It would be very useful to have a nice table showing how long each species can survive inundation. Alas, such data rarely exist. Examples like Table 1.2 are notable for their relative rarity in the scientific literature – and Table 1.2 is small: in the Amazon, there may be a thousand species of trees and little or no information on traits (Parolin 2009). We might turn to physiologists for some assistance, hoping to extract trait matrices from published studies, but physiologists are often constrained in the number of species that can be studied given the difficulty of measuring physiological data. There are thus

Table 1.2 *Flooding is a common filter in floodplains. Trees differ in their tolerance to flooding (Crawford 1982).*

Species	Survival time (yrs)
Quercus lyrata	3
Quercus nuttalii	3
Quercus phellos	2
Quercus nigra	2
Quercus palustris	2
Quercus macrocarpa	2
Acer saccharinum	2
Acer rubrum	2
Diospyros virginiana	2
Fraxinus pennsylvanica	2
Gleditsia triacanthos	2
Populus deltoides	2
Carya aquatica	2
Salix interior	2
Cephalanthus occidentalis	2
Nyssa aquatica	2
Taxodium distichum	2
Celtis laevigata	2
Quercus falcata	1
Acer negundo	0.5
Craetagus mollis	0.5
Platanus occidentalis	0.5
Pinus contorta	0.3

many papers on flood tolerance of *Alnus* and *Phragmites*, but none comparing them simultaneously to other species in the same pool. Hence, these individual measurements on just a few species are of limited value to community ecology. And even for these common species, there is disagreement as to the physiological trait that confers flood tolerance (Armstrong and Armstrong 2005). Hence, one cannot assume that we have, for most areas, a simple table of comparative flood tolerance of trees. Unless, of course, you want to rely on descriptive data, using the tried-and-true logic that species that grow in wet areas must be flood-tolerant, which, while likely true, is completely circular.

It is important to distinguish between rankings of tolerances (such as Ellenberg indicator values) as opposed to traits that can be directly measured. Here is where screening an entire group of species for one or more traits is such an important contribution. The study of seed germination characteristics of 403 species of the British flora, compiled by Grime and his co-workers (Grime et al. 1981), using standardized growing conditions, was an essential contribution and a model. Later work expanded the matrix to 67 traits, but in a smaller set of 43 plant species (Grime et al. 1997). Other selected examples include xylem characteristics in 480 species of trees (Choat et al. 2012) and a set of 24 traits in 99 European bird species (Renner and van Hoesel 2017). Trait databases for plants now include the European plant trait data set LEDA (Kleyer et al. 2008) and the global leaf trait database GLOPNET (Wright et al. 2004). The TRY Initiative began compiling plant trait data sets that were scattered across personal computers and journal supplementary material into a searchable global database that now has nearly 12 million records from over 279,000 plant taxa (Kattge et al. 2020)! We will return to the topic of traits in Chapter 4, where we distinguish between screenings and compilations of traits.

d: Dispersal

We include dispersal as a term, partly for completeness. Neither Major (1951) nor Mueller-Dombois and Ellenberg (1974) include dispersal in their formulations, but they do include time explicitly, which naturally invites consideration of dispersal. Moreover, organisms differ in their dispersal abilities, in which case dispersal traits could be included in the trait matrix. Dispersal is clearly an important process in community assembly, but dispersal limitation likely becomes more important with increasing spatial scale. In one view, community assembly is a goodness-of-fit problem: given the set of species in the pool, and their traits, what is the best set of species to tolerate a given set of filters? In some cases, we may find it convenient to imagine dispersal is the same for all species, even if we know it is not; it is a simplifying assumption.

Dispersal is also a topic that affects the size of a species pool. At the large scale, a species pool exists and can be delineated precisely because there are realistic limits to long-distance dispersal. Prior to globalization, the biota of Earth as a whole did not mix freely, and hence, for example, Wallace (1876) was able to divide the world into six zoological regions: the Palaearctic, Nearctic, Ethiopian, Oriental, Australian and Neotropical. In the same way, Takhtajan (1986) was able to divide the world into six floristic kingdoms and 35 floristic regions (you can see the map in Figure 4.2). Barriers to dispersal explain, in part, why fragments of the old continent of Gondwana are still different from the rest of the world, and even why fragments of Gondwana such as Australia and South America themselves have a different flora and fauna. Since most of the northern hemisphere falls into a single floristic kingdom (the Holarctic Kingdom) with just nine regions, it is easy for northern ecologists to underestimate just how different species pools can be in sub-equatorial regions of Earth. These regions exist largely because of limitations on long-distance dispersal. It is interesting to note that Wallace explicitly discusses dispersal and its limitations, in his words, "barriers to migration" (p. 6). He notes that flightless birds "such as the ostrich, cassowary, and apteryx, are in exactly the same position as mammalia as regards their means of dispersal" (p. 16). He also describes the role of floating rafts of vegetation in dispersing mammals (p. 15). Barriers to migration are an important cause of different biogeographic realms, and therefore help determine the species pool of any particular location. The entire problem of invasive species is driven by humans transporting species across what were formerly ecological boundaries.

Of course, natural long-distance dispersal does occur. Owing to its relative infrequency, naturalists have long been intrigued by examples of long-distance dispersal. The first plant to colonize the new volcanic island of Surtsey, in 1965, was the beach plant called Sea Rocket (*Cakile maritima*) (Magnússon et al. 2014). One might argue that this does not qualify as long-distance dispersal since the plant was colonizing within a single floristic region, the circumboreal. Consider the Cattle Egret (*Bubulcus ibis*), which in the 1930s "began one of the most dramatic and best documented avian range expansions occurring this [meaning the twentieth] century" (Arendt 1988), by crossing the Atlantic Ocean to colonize the New World. Such long-distance dispersal is a rare occurrence, however, and that is what allows us to delineate species pools for different regions. For example, when a tree falls in Algonquin Provincial Park in Ontario, or in Yellowstone National Park in Wyoming, we do not have to consider the possibility that tree ferns will establish in the resulting gap. Such species are not a part of the pool. The odds of them dispersing spores all the way from the Luquillo mountains of Puerto Rico are vanishingly small. That, as we shall see, is why we need to define a pool as a group of species with a reasonable probability of being able to enter a community at a particular site. What "reasonable" might mean is something we will return to later in Chapter 3.

We do need to mention one other aspect of dispersal at this point. Although we can indeed delineate a pool of species, **P**, it is difficult to avoid the reality that some are more likely than others to arrive at a particular location. The greater the flow of propagules, the greater the likelihood that a species will successfully establish. So, in this way, one could argue that traits that measure dispersal ability are an important part of the trait matrix. Plants with wind-dispersed seeds, for example, are well known for colonizing gaps in vegetation (e.g., Harper 1977, van der Pijl 1982). Many marine animal species similarly have planktonic larvae for long-distance dispersal (e.g., Jackson and Coates 1986, Frid 1989). In general, small propagules disperse long distances but many animal-dispersed seeds can be rather large. Many plants and some animals also have clonal growth forms, where there is small-scale movement of adults by clonal growth; this, however, is more of a trait for holding space rather than colonizing new patches of habitat (Williams 1975, Grime 1979). One exception may be aquatic plants, where vegetative structures like turions and even rhizomes can be dispersed long distances by moving water (Sculthorpe 1967). Most clonally spreading species, however, have another means for long-distance dispersal, such as wind-dispersed propagules or planktonic larvae.

If we decide to include dispersal traits, there is a confounding factor to consider. The number of propagules that are available for dispersal will be partly determined by the species composition of adjoining landscapes. Species that are already locally common may therefore be disproportionately represented in the stream of propagules. Here we might wonder if we are entering a zone of circular thinking, along the line that species i is likely to occur because species i is common. This is not circular reasoning so much as accepting that communities and landscapes have a kind of biological inertia. Those species most likely to disperse to a community are indeed generally those that are already the most common. An exception would of course be a common species that did not produce propagules. One way to assess this process in real communities is to create artificial patches and monitor rates of recolonization (Hartman 1988). Another is to experimentally manipulate dispersal (Shurin 2000, Ejrnæs et al. 2006). It is also possible to use transplanted individuals to measure rates of survival and growth in clearings (Bertness 1991, Geho et al. 2007). The degree to which community composition is controlled by lack of propagules ("dispersal limitation") remains an open question. Overall, the greater the efficiency of dispersal in the species, the less dispersal limitation there will be in a community. Thus, it may be important to determine the dispersal traits for members of the species pool.

In conclusion, we may find it convenient to assume that dispersal is not an important factor, and simply approach community assembly as a question of goodness-of-fit in sets of trait matrices. However, it may prove necessary to include dispersal traits in the trait matrix, and perhaps even abundance of those propagules in the surrounding landscape. If dispersal is limiting colonization, then time becomes important in the problem of assembly. We discuss dispersal in relation to species pools in more detail in Chapter 3.

t: Time

You cannot talk about communities without some mention of time. Time was included in Equation 1.1, while, more recently, *Plant Ecology* (Keddy 2017) has an entire chapter on the topic, with time scales ranging from 10^1 to $>10^6$ years. Still, our intention is to keep it to a minimum in this book. We could say that our focus in this book is primarily about space, rather than time. We assume that with regard to the central issues of community ecology, we can mostly ignore really long periods of time, at scales greater than 10^6 years that include continental drift and speciation. We are interested here only in shorter time scales (mostly 10^1 to 10^2 years). We do not deny that evolution occurs on ecological time scales. Evolution and bio-geography are clearly important. But for the sake of simplicity and devoting attention to our specific goal for the book, we treat the products of evolution and biogeography as essentially fixed quantities.

At the other extreme, we can also ignore many events that happen within relatively short time periods that occur over months and days. Of course, time is inherent in the action of environmental filters. It takes time for a drought to arrive, and one can measure how long it persists. It also takes time for an individual or population to perish from drought, and the time to death will depend both upon traits the organisms possess and the severity of the water deficit. Similarly, it takes time for flooding to kill plants, sometimes just a few days, sometimes a few years (Table 1.2). Having said this, we can, quite conveniently and reasonably, ignore much of this informa-tion, since the real point is simply this: is the organism still alive by the time we sample the community? We can, more or less reasonably, ignore most short-term events and focus our attention on those indi-viduals that survive long enough to be sampled. How long other individuals lingered, and the unhappy details of their demise, need not concern us.

This is important enough to deserve a restatement. We have set an upper limit on time scales that is short enough to reasonably ignore evolution, and a lower limit long enough to remove short-term physiological activity in organisms. What we are left with, in particular, is the time involved in events such as succession and dynamics, which is a venerable topic in ecology (e.g., Clements et al. 1929), and one that frequently occurs on time scales of about 10^2 years. For example, a review of old field succession in North America (Wright and Fridley 2010) found that most fields will be 50 percent covered by woody plants within 50 years, some as few as a decade. Here is where factors like rate of formation of gaps in the vegetation, and rate of dispersal of propagules, do fit within the framework of community assembly. And this is where data on traits are useful, such as whether the species has buried seeds, what stimulates germination and how quickly seeds arrive from adjoining forests. Even here, one could take a longer time view: succession in eastern North America, after all, typically leads to a tree-dominated landscape, so eventually the ability to tolerate competition from other trees, or the ability to regenerate in gaps, will come to predominate the rules for community assembly.

SUMMARY

Here is the central challenge of community ecology: if we have a specified location with a known species pool **P**, can we use knowledge of the species traits **T** and the filters **E** at that location to predict the species composition of the community **C**? Such a class of rules can be called assembly rules. As we have summarized above, assembly rules involve more than simply asking (once again) whether there is a pattern in a selected set of species at a selected location. We have given an outline of the important elements of this assembly problem above and summarize them in Figure 1.5.

The upper half of Figure 1.5 shows the logical structure of community ecology viewed through the lens of trait-based

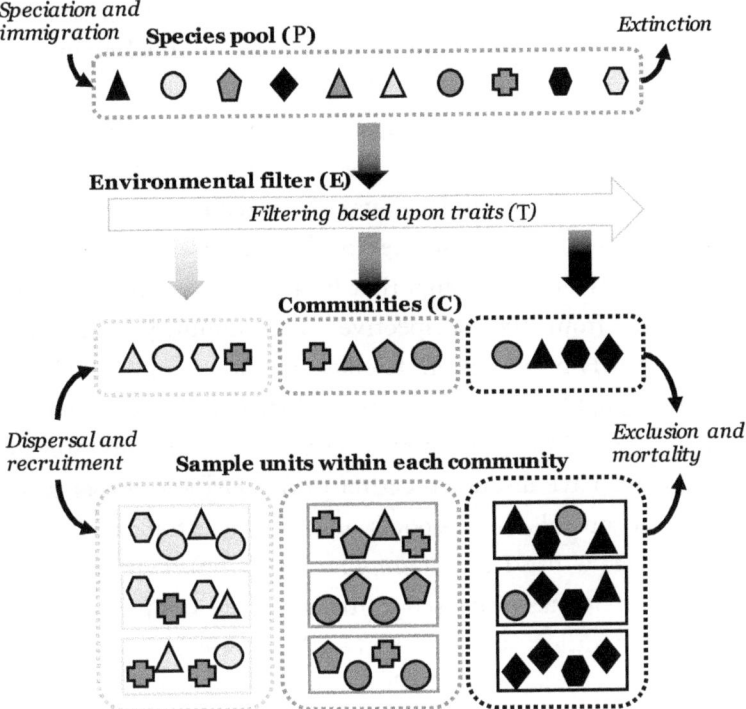

FIGURE 1.5 A general framework for community ecology. The species pool is the raw material. It increases in size from speciation and immigration and decreases in size from extinction (we discuss reasons why emigration is unlikely to affect pool size in the text). A small set of filters (E) determines which members of the species pool (P) occur in local communities (C). Some filters, like frost and drought, are physical factors. Other filters, like competition and predation, are biotic factors. Community membership increases with dispersal and successful recruitment, and decreases through exclusion and mortality. The actual sample units, shown below, usually reveal only a subset of the actual community.

community assembly. Species pools are filtered into habitats along environmental gradients based on their functional traits. In this book, we argue that this is the correct lens for community ecology, at least most of the time. In subsequent chapters we will explore the key elements of this figure, with particular attention to filters in

Chapter 2, pools in Chapter 3 and traits in Chapter 4. After laying the foundations, we describe quantitative translations of this framework into explanatory and predictive models in Chapters 5 and 7.

A careful examination of Figure 1.5 will reveal an asymmetry between the left and right side of the figure, between the factors that increase the pool on the left, and the factors that reduce it on the right. There are two inputs to the species pool. The first is speciation. Speciation happens slowly enough that P can be considered more or less a constant from the perspective of community assembly. The second input is immigration. It takes only a few individuals to establish a new population in a new ecoregion, so it is reasonable to assume that rates of immigration may be rather higher than rates of speciation. As we have discussed earlier in this chapter, dispersal is a key factor that controls species pools, and has been a topic of discussion since Wallace wrote *The Geographical Distribution of Animals* in 1876. Savile (1956) has discussed a broad array of examples of plant dispersal that potentially affected the North American species pool: plants that likely crossed the North Atlantic via Greenland, or the Pacific Ocean via Beringia "Whether they were carried by pack ice, blew across solid ice, were accidentally carried by birds or mammals, or were transferred in the mud layer of iced komatik runners is imma-terial" (p. 3437). As a result, many northern species have nearly cir-cumpolar distributions (compare this to the southern hemisphere in Figure 4.2, where barriers to immigration are wider).

Now, focusing on the right side of the figure, we have the counterbalancing processes. A symmetrical figure would have two elements on the upper right: extinction and emigration. Extinction is clear enough, and there is evidence for slow but steady rates of extinction through the fossil record, with occasional pulses of extinc-tion owing to dramatic geological events, such as the impacts of the asteroid that caused the extinction of the dinosaurs (Levin and King 2017). Also at the upper right, we come to the issue of whether there could be loss of species from the pool through emigration. Although it would be convenient, from the point of view of symmetry, to imagine that emigration in some way balances immigration, this is unlikely to be the case. There is one obvious reason: for the species pool to increase by one, it is necessary only that a few individuals arrive and successfully establish new populations. Exponential growth will take care of the rest. In contrast with long-distance dispersal, for the species pool to decrease by one, it is necessary for all the members of a population to leave an ecological region. Few examples come to mind. Even if we try to imagine a large population of migratory birds or mammals migrating to a new location, it seems unlikely that such a movement would completely remove a population from the species pool, but would rather leave behind stragglers. We cannot come up with a single recorded example of such an event. Cattle Egrets (*Bubulcus ibis*), for example, arrived in North America, but left a population behind in the Old World. In this sense, they immigrated to the New World, but did not emigrate from the Old World. Similarly, people may talk carelessly about species "migrating" north with climate change. The reality of the process is that during periods of warming climate, such as occurred at the end of the last ice age, individuals remaining at low latitudes mostly died, while individuals spreading to higher latitudes dispersed and established. The so-called retreat of species from the southern limits of their distribution is usually the result of death, not emigration, particularly in the case of sessile species like trees and corals. Hence, we do not include emigration as a process affecting species pools in Figure 1.5.

The lower half of Figure 1.5 exists mostly to remind us that other uncertainties will influence what we actually discover when we make observations on a particular piece of land. These are potential sources of confusion. Here are five of them:

(1) Any field sampling effort, even a morning excursion in the forest for bird watching, is constrained by the familiar pattern of species accumulation curves. Hence, observations of community composition are always affected by sampling effort (e.g., you may look ahead to Figures 3.8 and 3.9).

(2) The vagaries of dispersal mean that certain species may be absent not because they cannot survive in the habitat, but because they have not yet

arrived. This is an old theme going back to the days of Wallace (1876). We shall see in Chapter 3 how field experiments show that dispersal limitation is widespread in nature (e.g., Myers and Harms 2009).

(3) Local extinctions can occur from a wide range of phenomena, including biotic factors like competitive exclusion, as well as abiotic factors such as fire or flooding. Many, if not most, communities are in a state of recovery from the most recent natural disturbance (e.g., Connell 1978, Huston 1979). Hence, what we are observing is one small piece of a large-scale process of patch dynamics. These kinds of natural disturbances will appear several places in this book, particularly in Chapters 2, 4 and 8.

(4) The canonical distribution of commonness and abundance (you may look ahead to Figure 8.5) tells us that in any community, a small number of species will be common and a large number rare, without attributing this pattern to any specific cause. We may not see some species because their populations are just so small that they do not (yet) appear on the accumulation curve. We will have more to say on the canonical distribution in the final chapter.

(5) Finally, we are now living in an era where anthropogenic effects are further changing the state of natural communities, and patterns we see in our sample units may be distorted by new and relatively recent human impacts, be they mining, industrial-scale logging, or commercial overhunting. It is true that a bulldozer, skidder, trawler or herd of domesticated animals can indeed be treated as a kind of filter – and their effects may temporarily override all other factors that have been operating in a landscape for millennia. The lower part of Figure 1.5 is there to remind us that when we observe actual communities, we are frequently finding only a subset of the species that are expected based upon trait-based filtering. We raise these issues now because when we are exploring the basic rules of trait-based community assembly, we need to be respectfully aware of, but not distracted by, these phenomena.

It may help to think back to *The Theory of Island Biogeography* (MacArthur and Wilson 1967), where S was the number of species on an island. In this case, we are pursuing a vector **C** of length S in any specified piece of Earth's surface, including ponds or lakes, as well as real islands. More importantly, and unlike MacArthur and Wilson, we are explicitly after not just the number of species but the species list for each quadrat or island. In a perfect world, as we said at the beginning of the chapter, we would also want to know abundances of each species in the vector. Since the solution cannot be found on a species-by-species basis, we need to approach it through understanding the traits of the species, and the links between those traits and environmental filters.

We have spent some time on relatively basic topics to emphasize that part of the problem of community ecology has been lack of focus. Yes, ecosystems and communities and populations have frustrating complexities. And most certainly, if we look for complexity, we will find it. And, yes, of course, there are different scales at which C can be measured. And, yes, of course, multiple filters act simultaneously. And, yes, of course, there are rich natural historical details about dispersal and death. But we too often have allowed ourselves to be distracted by detail. Our objective in this chapter has been to introduce a simple logical structure underneath all the rich natural history of wild nature. Insights by scholars including Raunkiaer, Jenny, Major, Mueller-Dombois and Ellenberg all provide a formal structure for our thinking, but too often we have drifted back into detail rather than pursuing this logical structure to its conclusion, hence this book.

KEY POINTS OF THE CHAPTER

- There are four fundamental elements of community ecology: the species pool vector P, the local community vector C, a vector of environmental filters E and a vector of functional traits T.
- The central challenge of community ecology is whether, at a specified location with a known species pool, we can use knowledge of the species traits and the filters at that location to predict the species composition of the community. Many scholars throughout history have approached this challenge.

- Local communities (C) are subsets of the regional species pool (P) and have been filtered based on the matching of their traits to the local environmental conditions of the habitat.
- Dispersal, time and competition are also important factors in community assembly.
- One might argue that not only is the capacity to predict both the community vector C and changes in C with time one of the most intellectually challenging parts of community ecology, it is simultaneously one of the most significant moral challenges for protecting Earth's biota.

2 Filters

Filters as causal factors in communities. A first proposition. Drought.
Frost. Hypoxia. Salinity. Grazing. Wildfire. Predation. Competition.
Power. Two more propositions. Measuring the relative importance of
filters in field experiments.

THE POWER OF FILTERS

If we designate the community vector **C** as the main dependent variable for community ecology, then environmental filters **E** can be considered the independent variables. Filters are the environmental factors that prevent certain species in the pool from occurring in a specified community. That is, filters are the causes of the reduction from **P** to **C** in any specified area (recall Figures 1.1 and 1.5). Hence, they are of broad general importance. There are several somewhat redundant words in the language of community ecology that address the concept of a filter. It may be termed an **independent variable** that predicts the dependent variable **C**, the composition of a community. It may be termed a **causal factor** that produces the community **C**. More generally, it is often called simply an **environmental factor**. In certain cases, the filter can also be considered an **environmental driver** that produces specific changes in a community.

This proliferation of words does not really assist meeting our scientific goals, although each word may have a nuance that helps us to think about the challenges of community assembly. In this chapter we are going to start with practical examples of specific filters that have been well-studied, and which undoubtedly control the presence and abundance of species. We have selected this set of examples partly because of their widespread occurrence, partly because they illustrate clearly how factors remove species from pools, partly because they specify the traits that are involved and partly because they illuminate the process of community assembly.

We might at first be inclined to think that a very large number of filters is necessary to produce the subset S species from a pool of Q species. This attitude has in the past led to large numbers of studies that are poorly focused: "The effect of factor X on species Y ..." is an all-too-common title in ecological journals. Simply insert a factor (filter) and a species name and you have a grant proposal title, as well as a journal paper title, or book title if the species is of sufficient commercial importance. It is true that a large number of factors might be needed to explain the difference between **P** and **C** – with emphasis on that word "might." It is certainly true that if one picks potential filters more or less at random, many are also likely to be rather unimportant. So, we are going to consider the relative import-ance of factors further. We want to be able to rank factors in order of their importance, which we will frequently call their "power."

In fact, there was one other general criterion that has guided the selection of our examples: we wished to use environmental factors that are probably powerful filters. We suggest that the power of a filter can be measured as the proportional difference between **P** and **C**. The greater the difference, the more powerful the filter. It makes sense that in a world with large numbers of species, and potentially large num-bers of environmental filters, that we should focus our attention on those that are most powerful, which leads us to our first proposition in this book:

Proposition 1. In any habitat, the power of a filter can be measured as the proportion of species that it removes from the species pool.

Expressed quantitatively, the power of a filter can be esti-mated as

$$power = 1 - \left(\frac{S}{Q}\right),\tag{2.1}$$

where S is the number of species in the community and Q is the number of species in the pool. Weak filters will exhibit values closer to 0 and strong filters will exhibit values closer to 1.

We will look at real and quantitative examples of this proposition later in the chapter. Later in the book, we will consider how filtering occurs in multiple habitats across continuous gradients of environmental filters when we explore trait–environment relationships in Chapter 5. In general, any natural habitat fits somewhere along one or more intersecting gradients, so these gradients provide a natural context for any selected habitat. Gradients also provide a powerful tool for studying ecological communities (Keddy 1991a). At the same time, it is frequently useful to focus on one small section of a gradient, which we call a habitat in this book. Each gradient is associated with one or more filters. These filters are either abiotic or biotic factors that change along the gradient and create ecological communities by selecting a subset of species from the species pool that is suitable to a particular habitat. So, now it is time to look at some real examples of powerful filters. We start with filters involving water and oxygen. Since all life depends upon the presence of water, and most living organisms require oxygen, where better to begin?

DROUGHT ACTS AS A FILTER

Since all living organisms require water, drought is a powerful force. In the most general way, we can say that drought has been an organizing factor in ecological communities since the first plants and animals left the ocean some 450 million years ago. And, at the most general scale, the presence and distribution of deserts in particular, and drylands in general (Figure 2.1), shows how drought continues to shape communities globally. A quite specific set of traits is associated with drought tolerance in plants (Table 2.1) and this results in the characteristic growth forms that dominate such landscapes (Figure 2.2). Many, but not all, plants in drylands are succulents with reduced surface to volume ratios, reduced stomatal density, waxy coatings and the capacity to store water in stems. But not all are cacti. Since cacti evolved

Table 2.1 *Traits associated with plants that are tolerant of drought (Gibson and Nobel 1986, Archibold 1995, Bartlett et al. 2012, Choat et al. 2012).*

Relative importance	Trait
Primary	Reduced surface to volume ratio
	Leaf turgor loss point
	Xylem cavitation resistance
	Reduced density of stomata
	Spines and hairs for shading
	CAM photosynthesis
Secondary	Spines (defence against herbivores)
	Secondary metabolites (defence against herbivores)
	Camouflage (defence against herbivores)

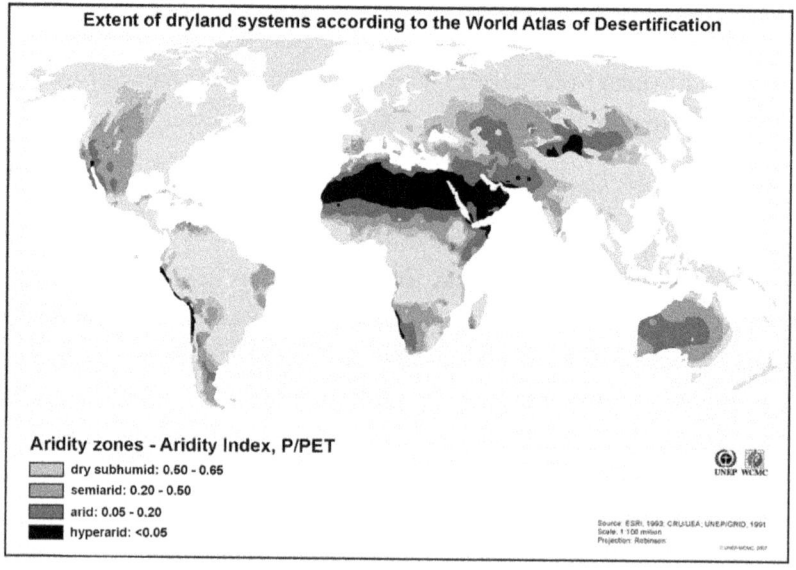

FIGURE 2.1 Drought is an important filter in many regions of the world. There are four levels of aridity: dry subhumid, semiarid, arid and hyperarid. (From Sörensen 2007, map 2, adapted from the *World Atlas of Desertification*, Cherlet et al. 2018.)

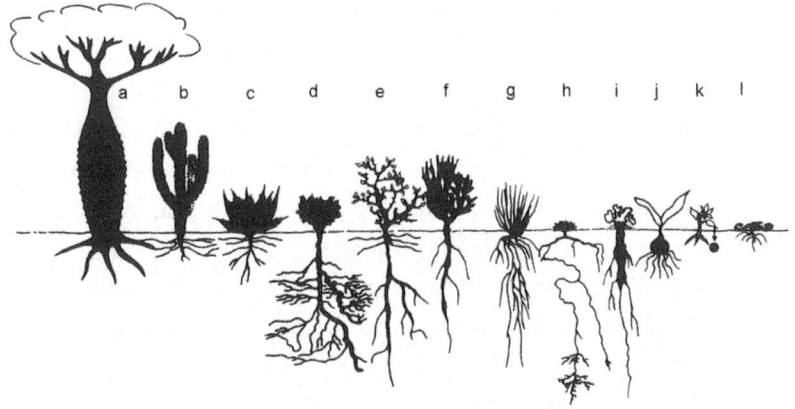

FIGURE 2.2 A small set of traits and growth forms occurs in desert communities. (a) Deciduous "bottle" trees with water-storing trunks (*Adansonia/Chorisa* type); (b) succulents storing water in the stem (cacti/ *Euphorbia* type); (c) succulents with water-storing leaves (*Agave/ Crassulaceae* type); (d) evergreen trees and shrubs with deep taproot systems (sclerophyll type); (e) deciduous, often thorny shrubs (*Capparis* type); (f) chlorophyllous-stemmed shrubs (*Retama* type); (g) tussock grasses with renewal buds protected by leaf sheaths, and with wide-ranging root systems (*Aristida* type); (h) cushion plants (*Anabasis* type); (i) geophytes with storage roots (*Citrullus* type); (j) bulb and tuber geophytes; (k) pluviotherophytes (annual plants); (l) desiccation-tolerant plants (poikilohydric type). (From Larcher 1995.)

in the New World, they are part of the species pool for New World deserts, but not those of the Old World, in which other plant taxa represent succulent life forms; examples include the families Euphorbiaceae, Liliaceae and Asclepiadaceae. In spite of the impression given by illustrations such as Figure 2.2, it is important to appreciate that many traits are continuous. This generalization includes not only physiological traits such as leaf turgor loss point (Bartlett et al. 2012) or xylem embolism resistance (Maherali et al. 2004, Choat et al. 2012), but also simple morphological traits such as surface to volume ratio or spine length.

This one filter, drought, affects more than 40 percent of the land surface of Earth. It not only shapes the flora and fauna of each community, but places important constraints on human uses of these regions.

It is also closely tied to contemporary concerns about possible changes in climate, anthropogenic and otherwise. In a larger context still, we note that the evolution of humans in Africa, and the dispersal of different human cultures within Africa, have been closely tied to effects of drought (Reader 1997). This illustrates our point, made above, that relatively few filters may be of overwhelming importance in the assembly of ecological communities.

Maps like Figure 2.1 emphasize the effects of global climate as a filter. In other cases, drought can arise because of much more local conditions. Let us look at a local example, where we have more information on communities and pools. This example comes from Ontario in eastern Canada (Catling and Brownell 1999a). In the south-central part of the province, one finds areas of rock barrens underlain by granite or gneiss. These ancient rocks erode slowly so the conditions are low in nutrients and acidic (Catling and Brownell 1999a). The soils are thin, owing in part to previous glaciation, but also because soil forms slowly, and recurring fires remove accumulations of organic matter. The undulating bed rock generates slopes and steep ridges from which water quickly drains. Hence, dry soils during the summer act as a filter in these locations. The species pool for this area is approximately 2,000, the number of species in the Ontario flora (Morton and Venn 1990). The local constraints are so severe that such habitats have only about 70 species (Catling and Brownell 1999a, table 24.1). Trees are particularly affected; often they are entirely absent, hence the term "barrens." There are only six species of trees that can tolerate these extreme conditions: Red Maple (*Acer rubrum*), Jack Pine (*Pinus banksiana*), Red Pine (*Pinus resinosa*), Trembling Aspen (*Populus tremuloides*), White Oak (*Quercus alba*) and Red Oak (*Quercus rubra*).

Let us use these data to return to Proposition 1, where we suggest that one measure of the power of a filter is the proportion of the pool that it removes from a community. In Ontario, the effects of shallow and acidic soil produce a community with just 70 out of 2,000 species of plants. If we apply our simple metric in Equation 2.1,

$1-(S/Q) = 1-(70/2000)$ equates to a filtering power of 0.965. Of course, it is not entirely clear whether we should regard this as being the effect of one, or possibly several interrelated filters. The ultimate cause is domes of hard bedrock. But this geology has several consequences: shallow soil, higher rates of run-off, and acidity. We can tease apart these causal factors somewhat by looking at nearby limestone barrens ("alvars"), which tend to be calcareous and flat. The absence of trees is again partly a result of shallow soil and midsummer drought, but being calcareous, these areas have a flora with a larger number of species. A cumulative list of plant species compiled across a set of seven such alvars gives 121 alvar species (Catling et al. 1975, table 2); a shorter list of 93 species are considered characteristic of alvars (Catling and Brownell 1999b, table 23.2). With a flora of 2000 species, this gives us a filter power of somewhere between 0.879 and 0.953. Owing to differences in the regions sampled, and the criteria used to construct the species lists, it is difficult to say more with confidence, except that the filters on limestone barrens are somewhat less powerful than those in acidic barrens. This example also illustrates a point we are going to return to in the next chapter: the criteria for designating species pools. We need good standard procedures for creating species pools; otherwise, we end up comparing species lists that were created using different criteria. Later in the book, we will describe a second way to quantify the power of a filter as the statistical relationship between filters and traits.

COLD ACTS AS A FILTER

In our survey of filters on ecological communities, cold temperature stands out as possibly being as important as drought. One of the distinctive features of Earth is that water can exist in three discrete states: solid, liquid and gas. The transition between liquid and solid is particularly important for living organisms, since cells cannot metabolize when water is solid. The process of freezing involves crystallization and changes in volume, which can damage organelles and the all-important cell membrane. Yet many parts of Earth, both higher latitudes and higher elevations, are regularly exposed to freezing temperatures. In boreal and arctic environments, organisms are exposed to prolonged periods below −40 °C and minimum temperatures below −60 °C. Plants that survive such conditions have a variety of anatomical and physiological features to avoid freezing or to tolerate its effects (Larcher 2003, Strimbeck et al. 2015). Perhaps the simplest evolutionary response is reduction in height so the plant remains buried beneath the snow, which takes us back a century to Raunkiaer's observations on plant meristems and climates. In the most extreme environments, trees (Phanerophytes *sensu* Raunkiaer) simply cannot grow. That is, the climate filters all tree species from the global pool, where S for trees is simply zero.

Plant scientists have spent many years trying to quantify the relationship between tree species distributions and climatic factors. As Woodward (1987) observed on the first page of his book *Climate and Plant Distribution*, "Any cursory glance at vegetation and climatic maps of the world cannot fail but to impress by the very close correspondence between the two distributions." The challenge is to make the link quantitative. It seems logical to begin with extreme conditions, since it is the extremes, not the means, that likely act as filters. There are multiple ways of using climatic data to quantify filters. One older but still relevant example (Ouellet and Sherk 1967) uses the monthly mean of daily minimum temperature as the predictor. In this case, the objective is to map plant hardiness zones. Each zone represents a further reduction in the pool of woody species that can tolerate the coldness of winter. Although this example may seem dated, it shows how traits and filters can be quantitively linked to make predictions of community composition, and so we will walk you through it in some detail. The equation used by Ouellet and Sherk to predict the number of tree species (Y) in a community was

$$Y = -67.62 + 1.734X_1 + 0.1868X_2 + 69.77X_3 + 1.256X_4 \\ + 0.006119X_5 + 22.37X_6 - 0.01832X_7 \qquad (2.2)$$

where:

X_1 = monthly mean of the daily minimum temperatures (°C) of the coldest month;

X_2 = mean frost-free period above 0 °C in days;

X_3 = amount of rainfall (R) from June to November, inclusive, in terms of $R/(R + a)$ where $a = 25.4$ if R is in millimeters and $a = 1$ if R is in inches;

X_4 = monthly mean of the daily maximum temperatures (°C) of the warmest month;

X_5 = winter factor expressed in terms of $(0 °C - X_1)R_{jan}$, where R_{jan} represents the rainfall in January expressed in millimeters;

X_6 = mean maximum snow depth in terms of $S/(S + a)$ where $a = 25.4$ if S is in millimeters and $a = 1$ if S is in inches; and

X_7 = maximum wind gust in (km/h) in 30 years.

The lower the value of Y, the fewer species of trees that can be found in a community. Looking more closely at this equation, the first term (X_1) is indeed set by the coldest period of the year. Some of the other terms include environmental conditions during the growing season, such as rainfall (X_3). The term X_6 actually includes the potential of snow to shelter at least some parts of the plant.

There is now a large scientific literature on plant hardiness zones, and entire websites as well. Figure 2.3 shows plant hardiness zones for the USA. Similar maps of plant hardiness zones are also available for other parts of the world. Such maps are obviously similar to maps of vegetation or ecological regions, but note that calculations for plant hardiness zones are predictive rather than descriptive.

HYPOXIA ACTS AS A FILTER

Many aquatic animals have their distribution and abundance partly controlled by the amount of dissolved oxygen in the water. Here are the basics. Most animals require oxygen for respiration. Although the atmosphere is nearly 21 percent oxygen, this gas dissolves into water at only low concentrations, typically measured in parts per million. Low oxygen levels are referred to as hypoxic conditions; the complete

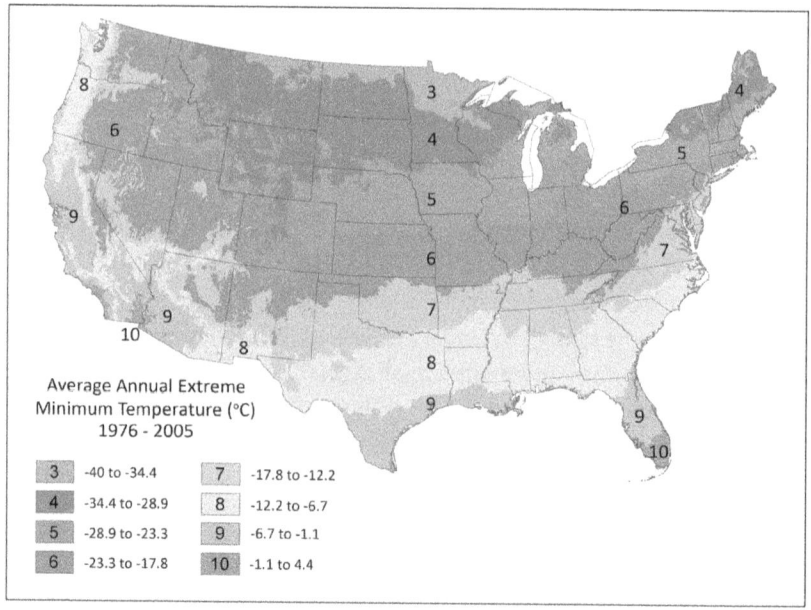

FIGURE 2.3 Plant hardiness zones based on average minimum temperature for the USA. (Produced by the US Department of Agriculture.)

absence of oxygen is anoxia. In colder climates, ice forms on the surface of lakes and ponds, forming a further barrier between atmospheric oxygen and water, a barrier that may last for months at a time. Hence, in cold climates the winter is typically the time when hypoxic conditions may kill aquatic life. However, warm water contains lower levels of oxygen than cold water. Hence, in warmer climates it is periods of higher temperatures that can lead to hypoxic conditions. In both climates, the presence of organic matter in the water can further reduce oxygen as microorganisms use dissolved oxygen in consuming the organic matter. Indeed, it is the abundance of organic matter that is associated with dead zones that occur in lakes and estuaries (Vallentyne 1974, Turner and Rabelais 2003). So, in general, low oxygen levels are an important filter for fish, amphibians, reptiles and aquatic invertebrates. Here we will look at the specific example of tropical fish in the Amazon River.

The Amazon basin has some of the largest areas of flooded forest in the world – some 70,000 km^2 – and the largest number of freshwater species in the world (Lowe-McConnell 1975, Goulding 1980). Some forests are flooded to depths of 15 metres and for up to 10 months of the year. Up to 3,000 species of fish may inhabit this region; of the more than 1,300 described to date, about 80 percent are either cat-fishes or characins. The latter group has radiated extensively in the Amazon lowlands, and includes carnivores, frugivores, detritivores and planktivores.

The annual cycle of tropical fish is closely tied to periods of inundation: "In both Africa and South America where much of the land is very flat peneplain, the rivers inundate immense areas, on a scale unknown in temperate regions. Submerged seasonally and drying out for part of the year, these floodplains are interspersed with creeks, pools and swamps, some of which retain water through-out the year" (Lowe-McConnell 1975, p. 90). Although rains occur in the summer, flood peaks occur well after the rains have started; the delay depends upon the origin of the main flood water and how long it takes to travel downstream. Rising water floods channels and creeks, releasing fishes imprisoned within ponds and swampy areas. Higher levels then create a vast sheet of water. This water is enriched in nutrients from decaying organic matter, including the droppings of grazing animals, perhaps first baked by the sun or fire. "This leads to an explosive growth of bacteria, algae and zooplankton, which in turn supports a rich fauna of aquatic insects and other invertebrates. The aquatic vegetation, both rooted and floating, grows very rapidly" (p. 92). Many fish then migrate upstream and move laterally onto the floodplain to spawn. The eggs hatch within a few days, so the young appear when food is plentiful. "The highwater time is the main feed-ing, growing and fattening season for nearly all species" (p. 93), but as nutrients are depleted and water levels fall, the fish move back into the main river. Some fish are killed by being stranded in drying pools, and predators often hunt the mouths of channels leading back to the main stream. Fish may mature within one or two years, and life cycles are generally short, so that catches can be closely connected to single good or bad spawning years. Some fish make long migrations between feeding and spawning areas; movements of 600–700 km in each direction have been recorded from the Parana and 125–400 km in the Niger. The same general sequence of events occurs in rivers throughout the tropics, including Africa, South America and Asia, although the timing of floods and the species involved may differ (Figure 2.4).

Hypoxic conditions pose a major constraint upon these fish, particularly after flooding, when shallow pools in floodplains are left filled with debris and exposed to the warming effects of sunlight (Junk 1984). The sorts of adaptations to hypoxia that can occur in aquatic animals are well illustrated by Amazonian fish found in such pools (Kramer et al. 1978, Junk 1984, Junk et al. 1997). Using closed respiration chambers, Junk et al. (1997) found that lethal oxygen concentrations were normally less than 0.5 mg L^{-1} O_2, although some species from well-oxygenated bays were sensitive to levels twice this high. So, as a first approximation, think of whether dissolved oxygen is greater or less than 1.0 mg L^{-1}.

Some 9 or 10 families of Amazonian fish include species that can remove oxygen directly from the atmosphere; the swim bladder of *Arapaima gigas* can act like a lung, and catfish may use the stomach to remove oxygen from swallowed air. Other fish can, within a few hours of hypoxic conditions, enlarge the lower lip to better extract oxygen from water. In one small floodplain lake, Junk (1984) reports that 40 out of 120 species could survive pronounced hypoxic conditions. Of these, 10 could take oxygen from the air and 10 could use the lower lip like a gill; the adaptations of the others were unknown.

Another way to escape hypoxia is to approach the surface and gulp fresh air. There are 347 species of fish in 49 families known to be air-breathing, and in nearly all, air-breathing is an adaptation to hypoxia (Graham 1997). One well-known example is the lungfish (six

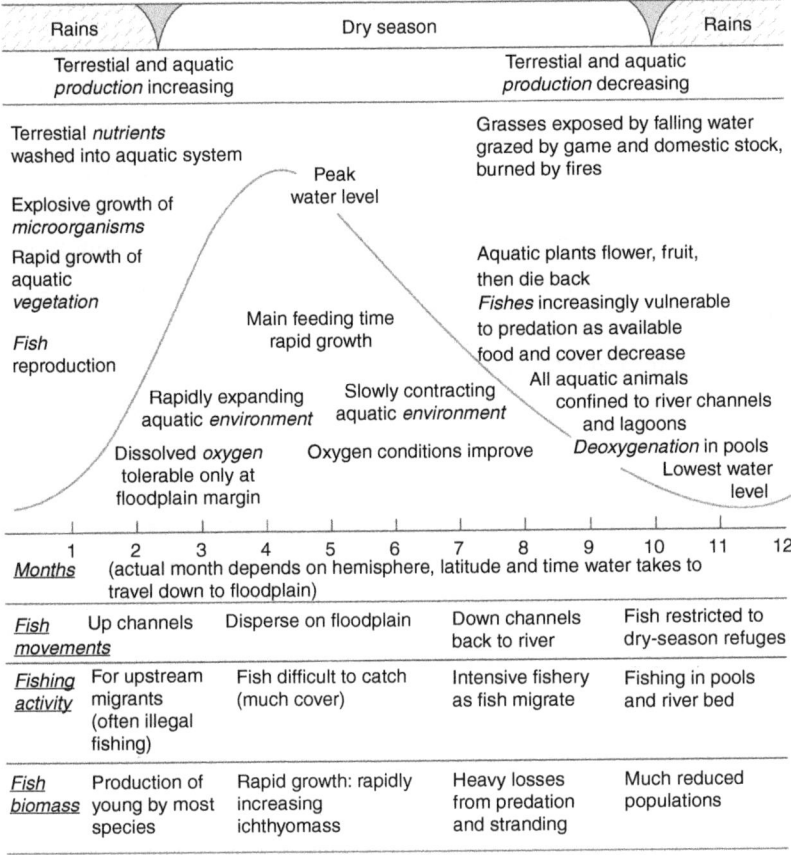

Rains	Dry season	Rains
Terrestial and aquatic *production* increasing	Terrestial and aquatic *production* decreasing	

Terrestial *nutrients* washed into aquatic system

Explosive growth of *microorganisms*

Rapid growth of aquatic *vegetation*

Fish reproduction

Peak water level

Main feeding time rapid growth

Rapidly expanding aquatic *environment*   Slowly contracting aquatic *environment*

Dissolved *oxygen* tolerable only at floodplain margin   Oxygen conditions improve

Grasses exposed by falling water grazed by game and domestic stock, burned by fires

Aquatic plants flower, fruit, then die back
Fishes increasingly vulnerable to predation as available food and cover decrease
All aquatic animals confined to river channels and lagoons
Deoxygenation in pools
Lowest water level

Months	1	2	3	4	5	6	7	8	9	10	11	12

(actual month depends on hemisphere, latitude and time water takes to travel down to floodplain)

Fish movements	Up channels	Disperse on floodplain	Down channels back to river	Fish restricted to dry-season refuges
Fishing activity	For upstream migrants (often illegal fishing)	Fish difficult to catch (much cover)	Intensive fishery as fish migrate	Fishing in pools and river bed
Fish biomass	Production of young by most species	Rapid growth: rapidly increasing ichthyomass	Heavy losses from predation and stranding	Much reduced populations

FIGURE 2.4 The distribution and abundance of Amazonian freshwater fish is strongly affected by hypoxia. During annual high water levels, fish move into the floodplain to feed. As water recedes, many fish are trapped in pools of warm shallow water, where decaying organic matter further reduces oxygen levels (adapted from Lowe-McConnell 1975).

species, including *Neoceratodus forsteri* in Australia). Other examples are the bichar (*Polypterus* spp. in Africa), the gar (*Lepisosteus* spp. in North and Central America), and the bowfin (*Amia calva*), all of which are predators in shallow warm waters. In addition to breathing air, some can walk away from the shrinking pool. The walking catfish (10 genera and ca. 75 species, including *Clarias* spp.) are native to Asia and Africa, where they normally inhabit warm hypoxic water. When stranded, they can move across land, partly by wriggling, partly aided by their pectoral fins.

An alternative to walking away from a hypoxic and drying pool is to bury in the mud and wait for the next flood. Lungfish, which occur in South America, Africa and Australia, will construct mucus-lined burrows and breathe through their modified swim bladder until the site is re-flooded (Graham 1997). Fossil burrows containing lung-fish have been found as far back as the Permian era. Some types of walking catfish can also walk on land, while others are known to dredge pools to make the water deeper.

Returning to the Amazon, let us consider the fish species found in a floodplain near Manaus, near the confluence of the Solimões and Negro rivers (Anjos et al. 2008). The fish were collected from two sets of sites: five with low oxygen (hypoxic, with dissolved oxygen concentration 0.20–0.98 mg L^{-1}) and five with normal oxygen levels (normoxic, dissolved oxygen concentration 2.34–6.50 mg L^{-1}). Note that these lie below and above the reference level of 1.0 mg L^{-1}. A total of 1,174 fish were caught, representing 22 families and 103 species. In the hypoxic conditions there were 78 species, while in the normoxic conditions there were 81 species.

One way to interpret these data is to begin with the pool of 103 species, obtained by combining fish caught in all the samples, including the two habitat types. This would, of course, be a conservative estimate of the pool, since it does not include those species that may be present in the river that were not found in this study. However, it gives a first approximation, that low oxygen levels removed one-quarter of the species, giving hypoxia a power of 0.25 (i.e., 1 − (78/103)). This seems like a tidy story, with stressful conditions acting as a filter on fish communities. However, the story is more complicated, since the locations with normal dissolved oxygen were also missing almost as many species, apparently giving a power of 0.22 (i.e., 1 − (81/103)) for the more desirable oxygenated conditions! It seems most unlikely that normal oxygen conditions are a filter on fish survival.

This raises the topic of how biotic filters may simultaneously act upon species pools. In Chapter 1, we observed that some environmental factors are certainly more important than others, and that in most habitats it is likely that a small number of factors controls the species composition of communities. Hypoxia is such a factor. The above example suggests too that habitats with lower effects of abiotic filters (normal oxygen levels in the current example) may experience higher effects from biotic filters. The classical case is the expectation that competitive adversity will increase as environmental adversity decreases (recall Figure 1.4). However, competition is not the only possible biotic factor to be considered. While it is possible that fish communities in normal oxygen levels could indeed experience higher levels of competition, Anjos and his co-workers (Anjos et al. 2008) offer a different mechanism – predation. The habitats with normal oxygen levels have higher abundances of piscivores – that is, predatory fish (Figure 2.5). Thus, it is possible that hypoxic areas in the river serve as refugia from predation. And, if we accept this explanation, predatory fish removed nearly as many species from the pool as hypoxia, a particularly tidy (and probably fortuitous) trade-off between the effects of two filters. As this chapter progresses, we will see other cases where animals act as predatory filters on plants and other animals.

SALINITY ACTS AS A FILTER

We have just seen how hypoxia can affect fish communities in rivers. Hypoxia is also more generally important as a causal factor creating wetlands. Soil hypoxia is a secondary effect of flooding, and all plants that occur in wetlands must have traits for coping with this filter. However, we are going to set this topic aside, for the moment, since we return to wetlands in several other places in this book. For example, in Chapter 3 we explore how flooding affects species pools, and in Chapter 4 we explore the traits that allow plants to tolerate flooding. Here, we wish to make a general observation that flooding is another powerful filter, one that is probably as important as drought and cold. This is particularly true of habitats flooded by salt water. Consider just

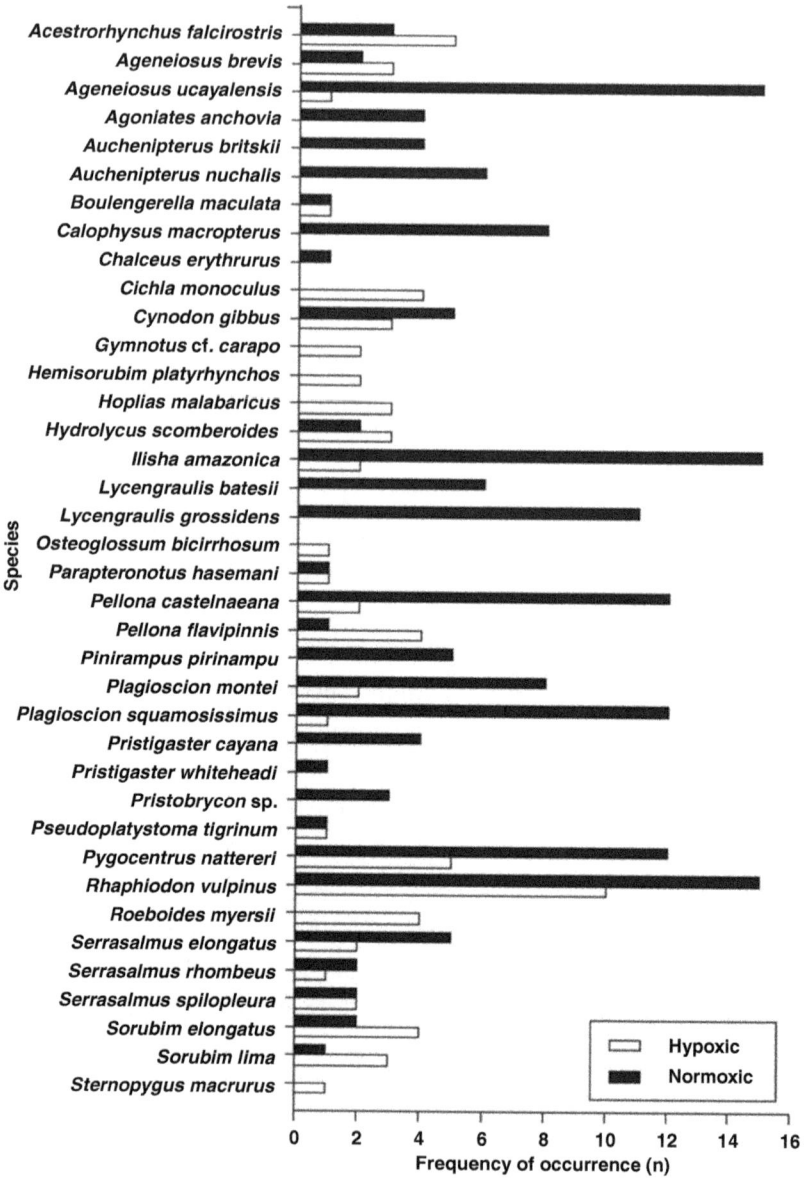

FIGURE 2.5 Predatory fish (piscivores) are more common in normally oxygenated water than in hypoxic water as shown by the frequency of occurrence of 38 species of piscivorous fish species in the Catalão region of the Amazon collected in 2003–2004. (From Anjos et al. 2008.)

one example as an appetizer: there are 2,600 species of palm trees in the world, but only four species occur in areas flooded with salt water, for a filter power of 0.999.

HERBIVORES ACT AS FILTERS

Herbivores have a powerful capacity to change species composition in plant communities (Keddy 2017). Examples include insects feeding on tree canopies, large mammals feeding on savannas and geese feeding in wetlands. In this section, we are not intending to address the general question of just when and where herbivory is most important; this varies with location, species and the method for measuring importance. We will focus here on some selected cases where herbivory has significant, even dramatic effects upon plant communities. Examples of strong and clear consequences for plants include effects of Muskrats on marsh vegetation (O'Neil 1949), effects of carp on aquatic vegetation (Badiou et al. 2011), effects of mammals on shrublands (Moolman and Cowling 1994), effects of White-Tailed Deer on forests (Latham et al. 2005) and effects of geese (Henry and Jeffries 2009) and snails (Silliman and Ziemann 2001) on coastal marshes. Indeed, in some cases, such as Muskrats, geese and snails, the herbivores may have such a strong effect that they not only cause major changes in composition, they completely remove the vegetation, leaving only bare substrate.

Goats are a particular problem, since their population is still growing globally, tracking to some extent the growth in human populations, and particularly affecting drylands (Keddy 2017, box 13.2). The world now has some half a billion goats, with a range far beyond their original location of domestication in the Middle East. And, since they are mostly domesticated, they are protected from natural population controls such as predators. There are well-documented cases of humans felling trees just to allow their goat herds a single meal (Thirgood 1981). Recurring grazing prevents trees from regenerating by seedlings, hence the example of deforestation of the Mediterranean, which is a likely precedent for many areas of Africa.

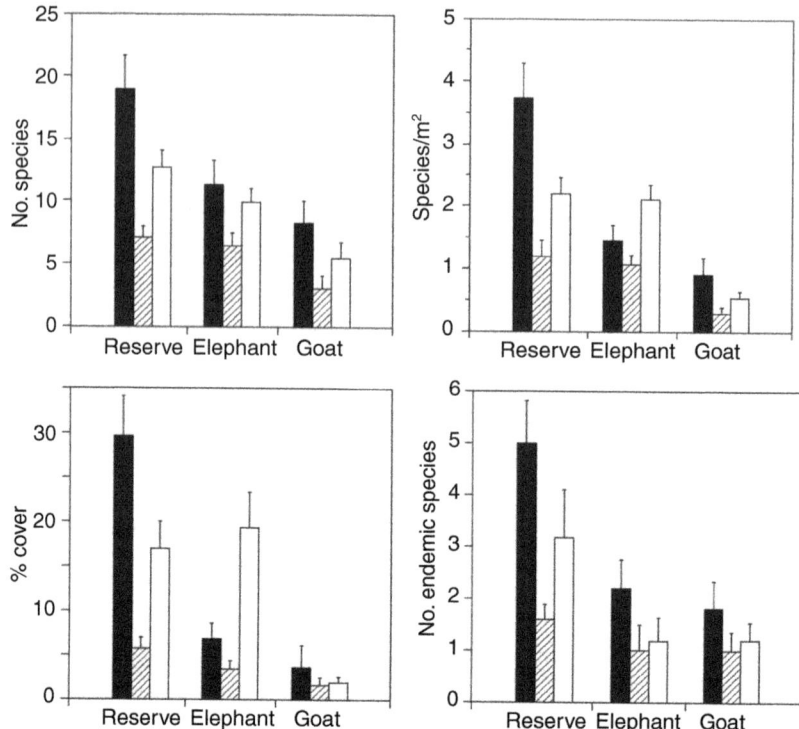

FIGURE 2.6 Grazing can be a powerful filter. Here are four characteristics of plant communities found in Addo Elephant National Park in South Africa. In each panel, the left-most cluster of bars shows control conditions, while the others show effects of elephant grazing and goat grazing. The three sub-groups in each cluster of bars show more detail: the three vegetation types are open areas, *Portulacaria* shrubs and *Euclea* shrubs. (From Moolman and Cowling 1994.)

Our example of grazing (Figure 2.6) therefore compares the effects of goats (an introduced herbivore) with the effects of elephants (the native herbivore) in the Cape Province of South Africa. The flora here has a large species pool including many succulents, and hundreds of endemic species in families such as Liliaceae, Asclepiadaceae, Crassulaceae, Euphorbiaceae and Mesembryanthemaceae. These landscapes have been grazed by elephants for millennia; goats, on the other hand, are introduced and their populations are expanding. Moolman and Cowling (1994) compared vegetation types among sites that are protected (with low grazing) versus sites grazed by elephants, and sites grazed by goats. Figure 2.6 shows that plant diversity (number of species and number of species per square meter) is strongly affected by grazing; that is, in the theme of this chapter, grazing acts as a filter. Second, grazing by goats is a stronger filter than grazing by elephants. Goats can remove nearly half of the plant species found in the reserve sites.

The enormous impacts of herbivores are particularly well documented on islands. Consider the island of Pinta, in the Galapagos (Hamann 1979, 1993). Pinta at 59 km^2 makes up only 0.75 percent of the total land surface of the archipelago; it is somewhat isolated and has an altitudinal range of 650 m. Because of the variation in elevation, the diversity of plants and animals is higher than that of the other Galapagos Islands: 180 taxa of higher plants are known, of which 59 are endemic to the Galapagos. In the past, grazing by tortoises was the primary source of herbivory. But the great tortoises on the island were driven to near-extinction in the 1800s, and then goats were introduced in the 1950s. The goats multiplied rapidly and soon began to cause deleterious changes in the vegetation. Closed forest and scrub were depleted, soil erosion started, and natural regeneration of several endemic plant species was prevented. The Galapagos National Park Service began a goat eradication campaign; the last goats were killed in 1990. Now biologists are monitoring the recovery of the plants. And, as an interesting aside, natural grazing by tortoises was restarted with 30 sterilized hybrid tortoises; there is still hope that a more genetically similar tortoise more typical of the original species may yet be found or bred (Hansen et al. 2010). So, over 200 years the island went through four phases, each defined by a filter: natural grazing by tortoises, expansion of woody vegetation (and hence shade) in the absence of tortoises, then severe overgrazing by goats, and now recovery from goat grazing with light grazing by reintroduced tortoises. Plant diversity and forest cover changed in each phase. Given the understanding of negative effects by introduced goats, and the development of new approaches to exterminate them, goat removal is becoming a powerful tool in restoration ecology. For example, a much larger island Santiago (>58,000 ha) was cleared of goats (>79,000 of them!) between 2001 and 2006, and the vegetation is showing rapid recovery (Cruz et al. 2009).

WILDFIRES ACT AS FILTERS

Wildfire is another disturbance agent that acts as a strong filter globally. Just like herbivores, fires can remove significant amounts of biomass from ecological communities. Fire is therefore an important kind of natural abiotic disturbance in communities, along with other disturbances including storms, landslides and volcanic eruptions (Keddy 2017). Fire selects for a different set of traits than herbivores, which are generally regarded as being more selective in removing biomass. At the same time, fire does not affect all species equally, since its effects are dependent upon traits such as bark thickness, flammability, resin content and resprouting, a topic to which we will return in the next chapter on traits. Fire not only removes biomass, but shapes the global distribution of entire biomes. Overall, fire is one of the major filters that controls the balance between tree-dominated and grass-dominated vegetation. Indeed, fire likely contributed to the origin of grasslands some seven million years ago in the Tertiary (Axelrod 1985). Because of its broad impacts, fire has been likened to a "global herbivore" (Bond and Keeley 2005). Fire is also closely linked to drought, as illustrated by the impacts of large fires in western North America during 2020 while we were writing this book, fires that may be related in part to an emerging North American megadrought (Williams et al. 2020).

How much biomass does fire remove from the planet? This is a difficult question to answer definitively. Bond et al. (2005) used a Dynamic Global Vegetation Model (DGVM) to simulate the distribution of closed-canopy forests, and then compared the vegetation patterns predicted by the model, with and without fire. It turns out that closed-canopy forests, which cover approximately one-quarter of the land surface of the planet, more than doubled in global extent when fire was turned off in the model. At more local scales, the impacts of fire upon the composition of communities have been documented in a broad array of natural communities, including prairies (Daubenmire 1968), pine savannas (Peet and Allard 1993), shrublands (Biswell 1974), montane and subalpine conifer forests (Heinselman 1973, Laughlin and Fulé 2008) and even wetlands (Loveless 1959). Regarding the Everglades, for example, Loveless (1959) concluded "The importance of fire and its influence on the vegetation of the Everglades can hardly be over-emphasized" (p. 7). We will have more to say about the effects of fire in later chapters. Now let us return to predators as important filters.

PREDATORS AS FILTERS IN FROG COMMUNITIES

There is a large scientific literature on the effects of predators upon prey. Here, we want to focus upon one subset of these studies: the effects of predators not upon single species of prey, but upon the composition of ecological communities. We have already encountered predatory fish in the Amazon. Another well-documented example is the effect of predatory fish upon amphibian communities. Amphibians often breed in ephemeral ponds, also known as vernal pools. One of the significant causes of mortality for young amphibians is death by desiccation when the pond dries up early. This would seem to be a good reason to not lay eggs in temporary pools of water. The principal advantage to ephemeral ponds is that the annual period of desiccation kills fish which would otherwise feed upon the eggs and larvae (Wilbur 1984, Snodgrass et al. 2000). One survey of 178 ponds in southern Ontario (Hecnar and M'Closkey 1997) noted the presence of all amphibians in ponds of three classes; our primary interest is those without fish compared to those with predatory fish. There was, however, a third class of pond: those with non-predatory fish. Figure 2.7 shows that the presence of predatory fish is associated with a dramatic decline in the number of amphibian species occurring in a pond. For a closer inspection of the patterns, Table 2.2 shows the abundance of

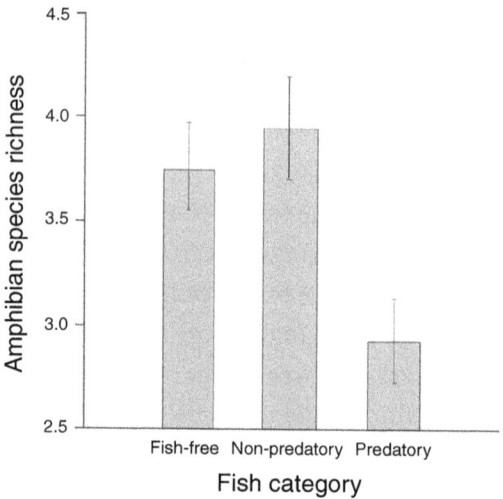

FIGURE 2.7 The number of species of amphibians in ponds can be greatly reduced by the presence of predatory fish. $n = 178$ ponds sampled between 1992 and 1994 in southwestern Ontario. (After Hecnar and M'Closkey 1997.)

common frogs in each of the three types of pond. There is a suggestion that the smaller species of frogs near the top of the table are more affected by predators. Some of the larger frogs may derive other benefits from being in larger ponds: the Green Frog (*Rana clamitans*), the largest in the table, overwinters in the tadpole stage, and hence cannot reproduce in vernal ponds. This frog species was the only one overrepresented in the larger ponds containing predatory fish. A similar pattern could be seen in bullfrogs (*R. catesbeiana*), uncommon in these ponds overall, and hence not shown in Table 2.2, but by far the largest frog in the fauna (12.1 cm). The American Toad (*Bufo americanus*) is known to have toxic eggs, larvae and adults. Generally, the ability to reproduce in the ponds with fish is likely associated with traits including (1) chemical repellents in larvae, (2) larger clutch sizes, (3) larger larvae and (4) larger adults (Hecnar and M'Closkey 1997). In some cases, frogs will not oviposit in water in which they detect predator fish.

The impacts of predatory fish on the abundance of individual species of amphibians can also be dramatic. Vredenburg (2004)

Table 2.2 *The frequency of seven common frog species among 178 ponds in southern Ontario, sorted into three groups of ponds: without fish, with non-predatory fish and with predatory fish. For each frog species, body length is given in centimeters. Prominent species of predatory fish included sunfish (65 ponds), bass (40 ponds), trout (20 ponds) and pike (10 ponds) (Hecnar and M'Closkey 1997).*

| Frog species | Pond status | | |
	None	Non-predatory	Predatory
Pseudacris triseriata (2.9)	16	1	1
Hyla versicolor (4.2)	16	14	5
Rana sylvatica (5.3)	16	8	6
Pseudacris crucifer (2.6)	38	27	14
Bufo americanus (7.1)	50	33	43
Rana pipiens (7.1)	53	35	32
Rana clamitans (7.4)	61	41	54

compared the abundance of Yellow-Legged Frogs (*Rana mucosa*) in a set of lakes in the Sierra Nevada. Yellow-Legged Frogs have been declining, and the disjunct southern populations are listed federally as endangered. In lakes lacking fish there were 700 tadpoles per 10 m of shoreline; this fell to fewer than 5 in lakes with trout. Caged trout were observed feeding on tadpoles. When fish were experimentally removed with gill-netting, frog populations recovered, in some cases after just one year of fish removal. To appreciate the significance of such observations, these trout have been introduced to previously fishless lakes for recreational fishing; of 50 lakes surveyed in 1996, 20 had introduced trout. In the western USA, thousands of lakes are stocked with trout, even in areas designated as wilderness.

PREDATION AS A FILTER

So far, we have mentioned two examples of predation as a filter: fish communities of the Amazon and amphibian communities in the temperate zone wetlands. There are good grounds for thinking that predation may be far more important than we realize. Here is the basic view: for the past century, the period over which community ecology has developed, we have been studying a world severely deficient in large predators. When Colinvaux wrote his book *Why Big Fierce Animals Are Rare*, which we mentioned in Chapter 1, the important general point we were noting is that big fierce animals are naturally rare, even in relatively intact ecological communities. This, along with work by Preston (1962), gives us a good general principle for understanding the relative abundance of species.

But modern ecology has not been based on the study of intact communities and ecosystems. There is a second reason why big fierce animals are rare: humans. Hence, we are also in an era, where to paraphrase Colinvaux, big fierce animals are unnaturally rare. For several centuries, large predators have been removed from wild nature around the world. On land, large predators like wolves and cougars and lions have been systematically hunted and poisoned, to the point that they are extirpated from large areas and considered endangered in many others (Ripple et al. 2014). In the ocean, large predatory fish have been harvested and turned into fish sticks and sushi: Myers and Worm (2003) estimated that the oceans had lost 90 percent of large predatory fish. And, in between the land and the water, in wetlands, not only wolves and cougars have been lost, but other large predators like alligators were hunted to near-extinction for meat and leather. Here is the general point: whether you are studying terrestrial systems, aquatic systems or their wetland interfaces, you have been studying ecosystems and communities in which large predators are mostly missing. One can even make the case that we have overemphasized the impacts of eutrophication in lakes and coastal zones because the large predators are no longer exerting top-down control of the system: Heck and Valentine (2007) summarized the data on benthic ecosystems, and showed that the effects of herbivores are significantly greater than nutrients, across a wide set of published studies.

This takes us into the issue of bottom-up versus top-down control (e.g., Power 1992, Sinclair et al. 2000). We will resist the temptation to explore this topic further, except where it is directly relevant to the topic of filters. Here are the general points. (1) Predation may be more important as a filter than most of us realize because we are living at a time when large predators are unnaturally rare. Moreover, (2) most of our published studies come from systems that are currently deficient in large predators, in which case they may also be experiencing unnaturally high densities of herbivores. And, parenthetically, since we are going to look next at experimental manipulation of filters, (3) large predators may be among the most difficult filters to experimentally manipulate.

HOW MANY FILTERS DO WE NEED?

We have now looked at a small set of powerful filters. Just how many of them do we need to understand the relative abundance of species in a particular community? There is no obvious answer, but contemplate the upper extreme. Let us consider a small lake, to set the foundations for our discussion of Frank Rigler in Chapter 4. Our lake is described by the vector $\mathbf{C}$ with S species (recall Figure 1.1), ranging from large lake trout (a top predator) to small phytoplankton. We could imagine a situation where each of the S species is independently controlled by the vector $\mathbf{E}$ consisting of f filters. Consider the many journal papers one can read with titles like "The effect of factors A and B on the ecology of species X at location Z." Usually such papers deal with a couple of factors, but the discussion then suggests many more could be involved, with the usual conclusion that more research is needed. In this worldview, the number of causal relationships we need to quantify in the lake is at least the number of species, multiplied by the number of possible causal factors, $S \times f$. This alone will keep many ecologists employed nearly *ad infinitum*. If we add in the possibility of pairwise interactions among species also being affected by environmental filters, then we have roughly $[S \times (S - 1)/2] \times f$, an even larger number, in which case we may safely say that we will never be able to understand even a single lake or single mountain on Earth. We return to this problem in the chapter on traits, and offer one prescription to solve the evident impossibility of understanding even one lake.

Fortunately, we think the problem is not as difficult as it seems. We suggest that there is ample evidence that the number of filters involved is actually quite small in most habitats. We also think there is ample evidence that many species in a community are controlled by the same filter. Putting actual numbers on this is difficult, but we think that the earlier examples in this chapter illustrate that drought in drylands is a powerful filter, and that other filters acting therein are secondary. Further, we consider it likely that a large number of the plants that do not occur in deserts are controlled by the same filter: lack of water. Hence, we offer two further propositions:

Proposition 2. In any habitat, only a small number of filters are likely to be important.

Proposition 3. In any region, a large number of species is likely to be controlled by the same filters.

Thus, the very large numbers of $S \times f$ or $[S \times (S - 1)/2] \times f$ are extreme upper limits on the complexity of wild nature. It is often far simpler than this. While nature is complex, it is not that complex. The species-by-species and factor-by-factor approaches tend to confuse our thinking and magnify the scope of the problem. How few factors are enough? What proportion of species are controlled by the same causal factor? These are important questions that we cannot adequately answer. A standard approach in multivariate studies is to measure as many factors as possible within budget, and then let statistics and model selection sort out which of the measured factors is most important. However, it would be useful to have some guidelines, and some rankings of the importance of filters. Although generalizations can easily become dangerous over-simplifications, generalization is one of the tasks for ecologists.

Consider the example of wetlands. There are huge numbers of studies on factors affecting particular species in particular wetlands.

Large conference proceedings are filled with individual case studies. What generalizations might we be able to draw that transcend species and location? We have seen above that hypoxia is a powerful factor controlling the abundance of fish in fresh water. (And, we note in passing, that if it is true of the Amazon, it is even more true of lakes at high latitudes and altitudes where the surface of the water is covered by ice for many months of the year.) Hypoxia is a key filter in all flooded habitats, acting on a wide array of different species. The book *Wetland Ecology: Principles and Conservation* (Keddy 2010) explicitly set out to explore the key causal factors in wetlands, at one chapter per causal factor. This may provide an example of how we might apply the above propositions. *Wetland Ecology* has chapters on causal factors: *Flooding, Fertility, Disturbance, Competition, Herbivory* and *Burial*. One widespread habitat, just six causal factors. Moreover, these factors act on many species simultaneously; that is, fish, frogs, turtles and plants are all controlled by the same set of factors. One could quibble that the list does not include predation, which is important for frogs, as described above. One could suggest that the chapter on herbivory should be expanded to include predation, since herbivory is just a kind of predation in which the victim is a plant rather than an animal. There is some logic to this argument, since top-down control of wetlands by predators is important in some cases; alligators may play an essential role in indirectly determining the composition of wetlands by controlling the abundance of herbivores like Muskrats and Nutria. From this perspective, alligators are a filter, and alligator hunting is removing a vital filter and perhaps contributing to the loss of wetlands along the Gulf Coast (Keddy et al. 2009). So yes, the list of causal factors might be extended to seven. And, for those who like exceptional cases, we note that the book has an eighth chapter, conveniently titled *Other Factors*. Yes, there are more than just six or seven filters, and in certain locations other filters like roads need to be considered. But the list is finite. More important, perhaps, is the relative power of each filter: Table 2.3 offers the opinion that flooding (hydrology) likely accounts for half of the variation

Table 2.3 *The estimated relative importance of filters ("environmental factors") that determine the properties of wetlands. These can be considered the key filters for assembling wetlands from species pools (Keddy 2010, table 12.2).*

Environmental factor	Relative importance (%)
Hydrology	50
Fertility	15
Salinity	15
Disturbance	15
Competition	<5
Grazing	<5
Burial	<5

seen in wetlands. Burial, on the other hand, likely accounts for less than 5 percent. Now, these are just estimates by one scholar, but they do tend to offer support for the first (and second) propositions.

A careful look at Table 2.3 suggests further simplification is possible. Grazing is just one kind of natural disturbance, so from one way of thinking, it could be rolled into that heading, reducing the number of filters by one. Competition and burial are less important, unless you consider them both to be elements of succession, which we know to be important in a wide array of wetlands. And why is succession not on this list of causal factors? But we now digress (although to answer that particular question, it is that rates of succession, and reversals of succession, are mostly set by flooding, fertility and natural disturbance, so in fact this causal factor is actually covered). Using this logic, Keddy (1983) suggested that only three causal factors are sufficient for a primary understanding of wetland ecology, with, of course, the proviso that other factors are undoubtedly also important in certain locations and at certain times. The important point from the objectives of this chapter is the evidence that Propositions 2 and 3 are illustrated by the way filters operate in wetlands.

MORE ON EXPERIMENTS AND FILTERS

The scientific literature in ecology has, over the last century, collected information on many different species, habitats and patterns. Often such studies include reference to factors that are acting as causal factors in general, and, in our context, filters. How might we move forward and use existing information to explicitly quantify and rank ecological filters? One option we have just used is to trust a professional opinion, based on field experience and wide reading. But this does have its limitations. How might we get beyond expert (or not so expert) opinion?

We suggest that three criteria stand out in importance. The first criterion is the requirement for experimental manipulation. There are a great many ecological studies that describe communities and then infer possible environmental causes. If we focus on studies that have actually manipulated environmental factors, rather than inferred them, the number of studies is much smaller. Moreover, many published studies report on only a single species. We can therefore add a second criterion, that studies of filters need to use multiple species simultaneously. The more species that are used, the greater the potential for generalization (which leads us to the topic of screening in Chapter 4). A third criterion is the need for multiple filters to be manipulated in the same study so that we may assess their relative importance. These three criteria significantly reduce the body of relevant scientific literature and guide future experiments. Of course, these criteria also significantly inflate the cost of such future research. If we add further criteria, such as experiments lasting multiple years, or with multiple treatment levels for each factor, the cost may become prohibitive. All the same, if we are going to be able to draw general conclusions about the relative importance of filters, large field experiments are going to be one of the tools.

There are growing numbers of studies that have manipulated multiple filters simultaneously and measured effects on multiple species: effects of fire, competition and grazing in the Serengeti (Belsky 1992), effects of ants and small mammals on Sonoran desert plant communities (Davidson et al. 1984), effects of disturbance, water and nutrients on old fields (Carson and Pickett 1990), effects of resources and herbivores on the understory vegetation of boreal forests (Dlott and Turkington 2000) and effects of fire, herbivory, disturbance and resources on coastal wetland communities in Louisiana (McFalls et al. 2010).

Let us look in more detail at one example that is very much in the news: the future of coastal wetlands, particularly those in coastal Louisiana. We will begin with a typical field experiment, and end with a discussion of filters and their implications for coastal restoration.

Our example comes from a large replicated field experiment that explored the effects of multiple factors (filters) acting on coastal wetlands near New Orleans (McFalls et al. 2010). These filters were chosen to represent the most important causal factors thought to be acting on these coastal wetlands, including sedimentation, nutrients, fire, herbivory and disturbance (Penfound and Hathaway 1938, Boesch et al. 1994). Biomass and richness were used as dependent variables to provide an aggregate measure of the relative effects of these filters upon plant communities. The results of the experiment showed that biomass of wetland plants increased monotonically with increased fertility (manipulated by both sediment and fertilizer), and decreased monotonically with increasing disturbance (manipulated by fire, herbivore exclusion cages and herbicide). There were multiple interactions with sediment addition. Similar, but less marked effects, were found for plant richness. Thus, it is possible to compare the effects of these filters by their impacts upon biomass (Figure 2.8).

One limitation of the above large study was lack of information on the effects of competition, and the way it might vary among species. Therefore, a sub-experiment examined the effects of three selected filters on 16 representative species of wetland plants (Geho

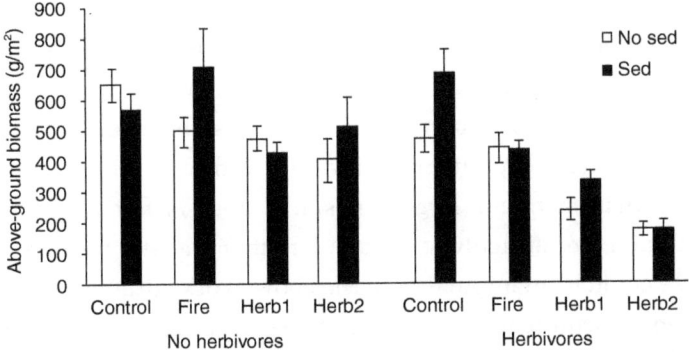

FIGURE 2.8 The relative importance of filters acting upon plant communities in coastal Louisiana as measured by above-ground biomass. Treatments included sediment addition, fire, herbivore exclosure cages and herbicide applications, single and multiple. There were significant effects from multiple filters, including disturbance, fertility and sediment. (From McFalls et al. 2010.)

et al. 2007). These species of plants were selected to represent an array of species and functional types typical of oligohaline marshes along the Louisiana coast, wetlands with a species pool somewhere between 40 species (saline marshes) and 200 species (freshwater wetlands) (Penfound and Hathaway 1938). The three treatments were: (1) an addition of 1 cm of sediment, (2) herbivore exclosures, and (3) removal of neighbours (no competition). Across 16 species, the experiment found that herbivory had the greatest effects upon species biomass, followed by competition. No effects of added sediment were found. There was also a significant interaction between competition and herbivory. Three species, including the original dominant tree (Baldcypress, *Taxodium distichum*), seemed able to survive only when herbivory and competition were simultaneously reduced.

A general conclusion from this pair of experiments is that biological filters are important in coastal wetlands. Although a great many studies (e.g., Penfound and Hathaway 1938, Boesch et al. 1994) describe relationships between flooding and vegetation, this does not prove that flooding is directly causing the observed changes in species composition. Rather, it may be that flooding is actually modifying biological filters such as competition and predation.

This pair of studies also illustrates the many challenges of designing field experiments to measure filters. First and foremost, filters that operate at large scales, like frost, or flooding, or alligator predation, are difficult to manipulate in a realistic fashion. None of these could be manipulated in the above work, although the manipulation of sedimentation was intended to alter flooding, indirectly, while exclosure cages simulated lower mammal grazing that might result from alligator populations. Second, many ecological experiments run only for a few years. This full experiment ran two growing seasons, with just one growing season for the sub-experiment on competition. Our experience was that the value of such long-term experiments was simply not appreciated, even in a coastal system that changes over decades and centuries. Third, there is the challenge of choosing, and creating, the correct filter levels. In this case, sediment was manipulated by adding 1 cm of natural sediment collected from another location. This was considered representative of sediment deposition from typical spring floods, but greater amounts of added sediment, or longer years of application, would certainly have changed the results, and likely would have increased the importance of sedimentation as a factor. At the same time, deep burial with sediment would be capable of killing all the plants (Keddy 2010). In conclusion, the experimental challenge is to manipulate the most important filters, apply the manipulations for many years, ensure that the manipulations use realistic levels for each filter, and to look at the responses of large numbers of species simultaneously.

Earlier in this chapter, we offered an encouraging proposition that only a small number of filters is likely important in any specified habitat. But which are the ones deserving most attention? What are the consequences of choosing the wrong ones in our models? Let us continue with two further examples from coastal Louisiana, looking at possible future scenarios for coastal wetlands (Figure 2.9).

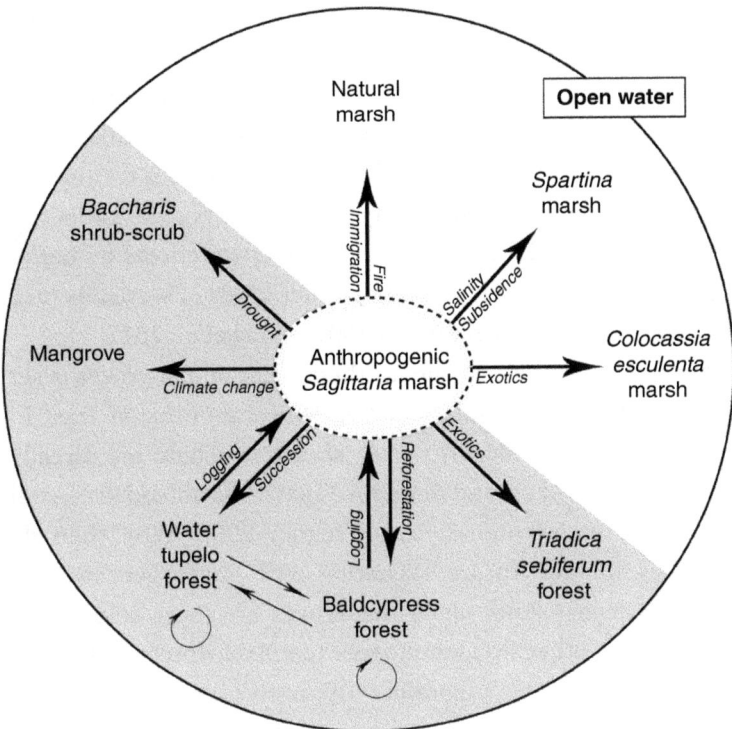

FIGURE 2.9 How many filters is enough? A coastal wetland (such as a *Sagittaria* marsh created by logging and saltwater intrusion) could change into many possible future states, depending upon the environmental filters that operate. For example, with fresh water, sediment and less herbivory, it might be possible to restore the site to its original species composition, Baldcypress swamp (bottom), while a slightly warmer climate might convert the community to mangrove swamp (left). (From Keddy et al. 2007).

The current paradigm in coastal restoration in Louisiana focuses on flooding and sedimentation as the dominant factors (e.g., Boesch et al. 1994, Visser et al. 2004). Both of these are physical, or abiotic, factors. Yet field experiments, such as those just mentioned, repeatedly show that biotic factors like herbivory are also important. For example, in order to re-establish coastal wetlands, it may be necessary to reduce herbivory, which will require larger populations of (larger) alligators to consume coastal herbivores. It is not just alligators, but *large* alligators that eat both native herbivores like Muskrats and introduced Nutria (Keddy et al. 2009). But large alligators are killed for their hides, in which case alligator hunting may be interfering with coastal restoration. Standard coastal models do not include this factor, and the effects of herbivores are mostly ignored in current restoration activity, except for programs which pay bounties on Nutria. Louisiana is not some sort of special case: there are top-down effects by herbivores upon plant communities in many other kinds of wetlands (van der Valk 1989, Silliman and Ziemann 2001, Wood et al. 2017).

Climate is also an issue. The northern limit of mangroves is set mostly by occasional frost pulses during winter months, at least in Florida (Stevens et al. 2006, Stuart et al. 2007). There are already thousands of hectares of coastal mangrove-marsh shrubland in coastal Louisiana (Louisiana Natural Heritage Program 2009). Why, then, do so few studies include future scenarios with mangrove swamps around New Orleans? And, since mangroves are more tolerant of salinity than many other floodplain trees like Baldcypress, it may be that mangroves would be a possible alternative restoration target instead of Baldcypress and tupelo swamps. Thus, a single filter, winter frost, could completely change coastal restoration scenarios for Louisiana in general and New Orleans in particular.

As these two examples show, then, the action of filters is not just an issue in theoretical community ecology, but has important consequences for management. By selectively choosing which subset of filters we study, and which subset we include in our models, we can produce very different scenarios for the kinds of communities that might arise. Hence, it is vital that we focus future work on the most important filters, even if they are also expensive and difficult to study.

CONCLUSION: THE UBIQUITY AND POWER OF FILTERS

In this chapter, we have seen seven main examples of causal factors that act as filters. Five were abiotic and two were biotic. We could continue to list additional causal factors, and collect interesting examples. But these examples are likely sufficient to illustrate the generality of the phenomenon we are describing. While there may be a large species pool of organisms that are potentially available to occupy a particular site, often a single filter can greatly reduce both the number of species actually found in the community and the abundance of each species.

KEY POINTS OF THE CHAPTER

- Environmental filters subset the community **C** from the species pool **P**.
- Proposition 1: In any habitat, the power of a filter can be measured as the proportion of species that it removes from the species pool.
- These filters can be abiotic or biotic in origin. Drought, frost, hypoxia, salinity and wildfires are important abiotic (physical) filters, while predation (including herbivory) and competition are important biotic filters.
- Proposition 2: In any habitat, only a small number of filters is likely to be important.
- Proposition 3: In any region, a large number of species is likely to be controlled by the same filters.
- The power of a filter can be measured as simply 1 minus the ratio of the number of species in the community to the number of species in the pool, $1 - (S/Q)$. It can also be estimated from field experiments.
- It is important to identify and rank the filters that control species composition.

3   Species Pools

Definition. Data availability. Kinds of pools. A general model. Immigration and local extinction. Examples. Experiments. Invasive species. Measuring Pools.

THE LIST OF PARTS

A list of parts is the obvious beginning for any problem of assembly, be it the assembly of a bookcase or the assembly of mammals in a national park. The list of parts for community assembly is the pool, which is defined as *the list of species which could occur in a particular ecological community* (e.g., Diamond 1975, Wiens 1983, Keddy 1992, Pärtel et al. 1996). The size of this pool is determined by speciation and extinction rates (recall Figure 1.5) and, locally, to a lesser degree, by immigration. Since rates of speciation and extinction are generally slow relative to rates of ecological processes that are driven by filters, we may treat the pool for a habitat as a more or less stable quantity. This may frustrate those who think in terms of evolution, but the failure to distinguish among rates of different processes is perhaps one reason for the confusion in some areas of ecology today. We must of course acknowledge that the pool of species on Earth has changed in numbers and composition through time (e.g., Niklas et al. 1983, 1985, Levin and King 2017). However, we currently occupy a specific point in time, in which case the changes in the past, and the prospects for the future, can be set aside, allowing us to treat the pool as a constant. With this simplifying assumption, the numbers of species that occur in each pool provide a body of data that is worthy of study in its own right.

We actually have a huge database of species pools for many kinds of organisms and many parts of the world, even if this database does not explicitly mention the word "pool." This huge database is the world collection of field guides and identification manuals describing the flora and fauna of different regions of Earth. To illustrate, consider Lanark County, where Paul lives. Species pools can be found in the following sources:

Birds: *A Field Guide to the Birds* (Peterson 1980)

Reptiles and amphibians: *Amphibians and Reptiles of the Great Lakes Region* (Harding 2006)

Plants: *A Checklist of the Flora of Ontario: Vascular Plants* (Morton and Venn 1990), which has been updated with new records to provide the *Southern Ontario Vascular Plants Species List* (Bradley 2013).

These books (Figure 3.1) are based upon vast numbers of observations by people who are not community ecologists, by and large. They have been verified by large numbers of naturalists and biologists who have used these guides to identify and enumerate the species. Another piece of good news is that the work is already done. That is to say, we don't have to start from scratch. We simply take the product off the shelf and use it. Now, if our budget is tight, we can buy a second-hand copy from a local used bookstore.

Of course, there are some complications (there always are).

FIGURE 3.1 Biologists have compiled guides to the plants, animals and fungi for many parts of the world, and these are a foundational source for information on species pools. These three guides are useful for Lanark County, Ontario, current home of the first author.

Consider the plants, which are the most speciose of the three groups given above. The flora of Ontario has more than 3,000 species, although the exact number depends upon interpretations of the fine distinction between species, subspecies and varieties. This large number partly reflects how speciose plants are, and how many habitats occur in Ontario. It also illustrates a problem with the issue of scale.

Ontario is a relatively large province and includes a range of habitats from deciduous forest in the south to boreal forest in the north, and even tundra (Figure 3.2). Hence, the checklist of the flora of Ontario includes plants from very different ecological regions. It is unlikely that tundra species are part of the pool for deciduous forests around the Great Lakes, just as it is unlikely that deciduous forest plants are part of the pool for the Hudson Bay lowlands. So this list of species for Ontario is actually bigger than we need. And, it illustrates a problem inherent in many guides – they are often based on political rather than ecological boundaries. So, we have to face a practical problem: many of our guides give us far larger pools than are necessary. This is not just a practical problem, but is also a conceptual problem as it raises the question of just how big an area is needed for a particular ecological study, and how boundaries should be drawn. We will return to this problem later in this chapter.

The same is true for *A Field Guide to the Birds*. It is true that this guide gives us a species pool for eastern North America, but if we were to use it in Lanark County, it would be misleading in the same way as the checklist of Ontario plants. Too many species. Indeed, one of the frustrations of using the *Birds of Eastern North America* in eastern Ontario is the large number of species that are illustrated but do not occur here. Consider the warblers on page 231. One of these is the Black-Throated Green Warbler, which is a common woodland bird here, and a likely candidate for a tract of mature forest. So, too, with the Parula Warbler, although it is more common in the boreal forest. However, immediately beneath it on the same page is the Prothonotary Warbler, which does not breed this far north. The Yellow-Throated Warbler, also on that page, breeds even further south. So, the actual

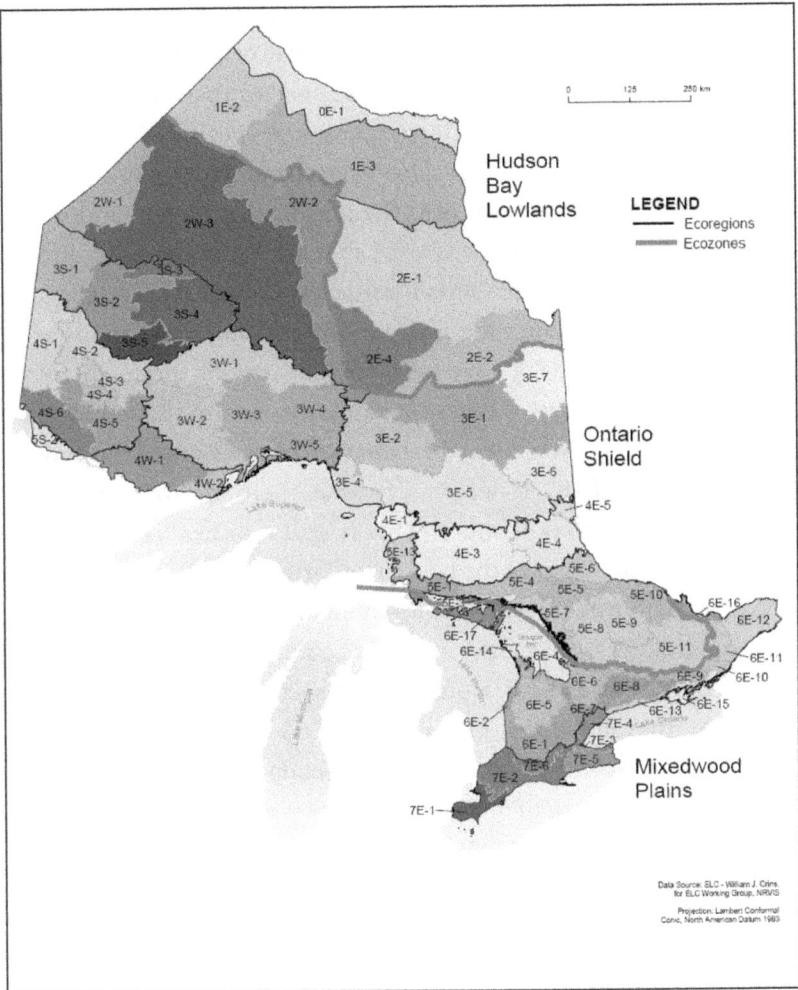

FIGURE 3.2 The species pool provided by the Ontario flora includes plants that occupy very different habitats, ranging from deciduous forests in the south to tundra in the Hudson Bay lowlands. This inclusion of many habitats complicates the search for species pools in published sources.

pool is half what we might gather from looking at this page in the guide. Here, again, the issue of scale matters: we need a list that includes only those species known to breed in this ecological region.

In the realm of reptiles and amphibians (Harding 2006), we fare somewhat better on this criterion, since instead of all of eastern North America we need only to focus on those species that breed near the Great Lakes. This reduces the large number of species from the pool for all of eastern North America. However, the problem is not entirely solved, since eastern Ontario is at the northeast extreme, and many species that occur south of the Great Lakes are not found here: no Box Turtles, no Bog Turtles, no Glass Lizards, no Timber Rattlesnakes.

Returning to plants, the pool for Ontario combines many ecological regions (Figure 3.2). We need smaller regions, preferably (but rarely) based on ecological boundaries. Many parts of the world have regional lists. Here, we have the Lanark County flora as an online database (White 2016). It informs us that there are 875 species of native plants in Lanark County. This is still a rather large number, but only about one-third of the pool in Ontario as a whole. We can also categorize them by ecological requirements, since another document (Oldham et al. 1995) assigns a wetness score for each species in the Ontario flora. For Lanark County, 390 species would be wetland plants (i.e., they have a wetness score of −3 to −5). So, for Lanark County, nearly half of the pool of plants occupies wetlands. Thus, as a first approximation, half of the flora occupies uplands, and half of the flora occupies flooded areas. We can also pull out individual groups, such as trees. There are 48 species of trees; of these, only 8 occur in wetlands.

The published list for Lanark County is therefore quite close to the true pool of species that might be expected to occur in local communities of Lanark County. Of course, the issues of scale remain. Less than 100 km to the south are additional species of trees, including Pitch Pine (*Pinus rigida*) and Chinquapin Oak (*Quercus muehlenbergii*). Perhaps we should consider these as candidates in the species pool. Actually, in this case it is unlikely, since extreme cold periods in the winter set a rather strict limit on trees in this part of the world. But the question needs to be asked: what about species not present, but nearby. And just how close is "nearby"? And, yet another question, how do we account for environmental change? How much change in climate will require us to consider such species a part of the pool that is simply lagging behind the warming climate?

There is also the issue of rare cases of long-distance dispersal. During November 2017, local naturalists were delighted to see a young (first year) Anna's Hummingbird in the town of Carleton Place in Lanark County, Ontario. This hummingbird is found on the extreme west coast of North America, particularly California and northern Mexico, about 4,000 km from here. Does this bird belong in the species pool for eastern Ontario? No, this is clearly an aberration. But it does illustrate how there need to be realistic criteria for compiling data on species pools, not just a list of everything that has ever been recorded in a particular region.

So, we end this section with questions and uncertainties, particularly over issues of scale. These uncertainties demonstrate why some care is needed in using existing species lists, and justify some of the theoretical inquires we raise later in the chapter. At the same time, we wish to emphasize the positive: thanks to several generations of explorers, systematists and field ecologists, we actually have rather good data sets for beginning our investigation of species pools. In this introduction, we may seem to have focused too much on eastern Ontario. We could have used any other part of the world, of course. We invite each reader to undertake a similar quest for the species pools that define your region. Meanwhile, eastern Ontario works quite well for introducing the availability of data, and the kinds of questions that arise.

TYPES OF POOLS

Thus far, we have avoided complications in nomenclature. For simplicity, we are using a relatively broad definition. The pool is the list of all species that could potentially be part of a community within a specified area. But there is actually a hierarchy of possible ways of defining pools bounded by two ends of a continuum.

At the smallest scale lies the species list found in a single quadrat, and ecologists call the length of this list "alpha diversity" (Pielou

1975). At the largest scale lies the list of species in the ecological region, the upper limit set by evolution and dispersal. Ecologists call the length of this list "gamma diversity" (Magurran and McGill 2011). In between lie a range of nested and overlapping possibilities. Existing studies on species pools tend to fall into two areas of concern: some focus on predicting alpha diversity and gamma diversity, while others focus more upon predicting composition (Zobel 2016). Our book focuses mostly upon the latter problem, predicting composition of ecological communities. If you can predict composition, then richness is also known, even if that number was not the original objective.

There is also a third type of diversity called "beta diversity." This quantity describes how much turnover in community composition exists within a landscape (or within your sample). Landscapes like the painting *Rambling Rio Grande* by Johnathan Harris, on the cover of this book, include riparian areas, arid woodlands, mesic mixed forests and alpine scree. This landscape exhibits higher beta diversity than a landscape with a single community. We mention this not because we will spend time pondering this number. Others have already done so in great depth (Anderson et al. 2011, Kraft et al. 2011, Myers et al. 2013). We mention this because understanding and predicting the turnover in community composition across environmental gradients is precisely the topic of this book.

Some authors use the word pool in a slightly different sense: they consider the pool to be only that list of species that could potentially occupy *a specific habitat type*. This use of the word is much narrower, and hence the number of species in the pool will be much smaller. What we are calling simply the species pool, Zobel (2016) called the *unfiltered pool*, in contrast with the *habitat-specific pool*. Here is a practical example that makes the distinction clear: continuing for a while longer with Lanark County, does the species pool for wetland communities include adjoining terrestrial species? In our use of language in this chapter, thus far, we say yes to that question. Terrestrial species may well disperse into wetland habitats, but they will not establish, and if they do, they will not survive. This is part of the process of filtering that we discussed in the preceding chapters. Hence, in this chapter, and in this book, when we use the word pool, we use it to mean the list of species that could potentially occupy the site before filtering begins.

In contrast, in some published studies, the focus is upon the habitat-specific pool. In this case, continuing with the Lanark County wetland example, one would not only eliminate from the pool all terrestrial species, but one might actually create narrower subdivisions within the wetland pool, such as species that occur mostly in calcareous fens, or peat bogs, or shallow water. In such cases, the list of species in the habitat-specific pool is often produced by adding up the species found in quadrats from such habitats – which is a methodological issue we discuss further below. Some authors use the word *regional species pool* instead of the term *unfiltered pool*. "A regional species pool comprises all species available to colonize a focal site" (Cornell and Harrison 2014, p. 45).

So, as you are reading the scientific literature on species pools, it is good to keep two distinctions in mind. Are people writing about unfiltered pools, or habitat-specific pools? Are they focused on richness or composition? Some of the confusion in terminology arises out of these differences.

Let us turn to an example from Estonia (Pärtel et al. 1996). The flora of Europe is well studied, in part because of the small number of species relative to the large number of botanists, and the tradition of botanical investigation going back to Linnaeus himself. Thus, we have accurate data on the regional species pool. There have also been some efforts to compile data on plant traits, including habitat requirements. Ellenberg and his co-workers (1991) compiled the list of habitat requirements for each species based upon the factors of, to name a few, light, soil moisture, pH and nitrogen, where each plant species has a score between 1 and 9 along each of these axes. For Europe, we know not only the pool, but also some general information on the ecological conditions in which each species in that pool is normally found. We can consider this list of Ellenberg scores to provide a kind of trait, with the proviso that it is not based upon screening under consistent conditions (see Chapter 4), but rather is a summary of years of field observations.

Pärtel et al. (1996) used the data in Ellenberg et al. (1991) to establish the list of species that could potentially occur in herbaceous vegetation in Estonia. The flora of Estonia contains 1,416 species in the pool, across all habitats. Pärtel and his co-workers focused on a set of 14 communities, including alvar, pine forest and raised bog. They knew which species occurred in their set of 14 communities from their field sampling. This gave them what they called the *actual species pool* (i.e., what we call the community vector **C**) for each of the 14 communities. This was determined first by adding up all the species from a set of quadrats, and then by adding other species that occurred in those habitats that had not been recorded in the set of quadrats (Pärtel et al. 1996, p. 114). By the time they finished the field work, they knew which species actually occurred in each of 14 locations. Of course, there was also judgment involved, as they had to set boundaries on the communities using various criteria. Anyone who has tried to collect a list of all the species in a habitat knows how much subjectivity can be involved. Do you include seedlings, or only adults? Do you include individuals that are common in an adjoining habitat? And, however much one tries to sample a homogeneous community, there are always, it seems, locations that seem different. Here is another example from Paul's forest in Lanark County. The landscape is underlain by gneiss, and the forest is therefore typified by species tolerant of acid soil, and therefore rather predictable and somewhat monotonous as judged by plant species composition. Yet, at one location, there is a single boulder made of marble, presumably carried there during the last ice age, and on this one boulder is a single individual of Bladder Fern (*Cystopteris bulbifera*). If you were sampling this forest, would you include this plant on the community list? Would you move your sample area? Would you consider it part of the species pool in a gneiss landscape, or an aberration? It is certainly part of the regional pool, and disperses enough spores widely enough to reach a single boulder within a gneiss landscape.

To determine their values for the *regional species pool* (i.e., what we call the habitat-specific pool), it was necessary to extrapolate. They defined their regional pool as "the set of species, occurring in a certain region (here: Estonia) which are capable of coexisting in a target community" (p. 111). Pärtel et al. (1996) first set out to define the habitat, or environmental conditions of each community, by calculating the mean Ellenberg indicator value for the species that already occurred there. That is, these factors were not directly measured, but inferred from the existing species and their indicator values. The regional pool was then constructed by finding all other species in the Estonian species pool whose indicator values fit with the average Ellenberg value for that community. Four Ellenberg factors were used: light, soil moisture, pH and nitrogen. The existing species in a habitat, then, were used as a kind of bioassay for the conditions of that habitat, and this bioassay guided the selection of other candidate species from the Estonian pool. Pärtel and his team used only 1,073 species of the flora for **P**, excluding groups with large numbers of microspecies (e.g., *Hieracium* spp.) and excluding species recorded only once or twice in Estonia. This was another judgment call. Then, not unlike the methods used by Raunkaier a century earlier (recall Chapter 1), they chose possible communities at random using what we now call Monte Carlo methods. By accumulating a large number of possible random communities, they were able to test whether the patterns they found were statistically significant.

Overall, they found a significant positive relationship between the actual species pool and the regional species pool (Figure 3.3a). They also found a significant positive relationship between the number of species in 1 m² quadrats and the actual species pool (Figure 3.3b).

Their work gives further insight into the problems of compiling data on species pools. When they first used the Ellenberg scores to create the species pool for each habitat, they concluded that "unrealistically small" pools resulted. The size of the pool was increased by using an amplitude of 1.5 relative units around the habitat mean.

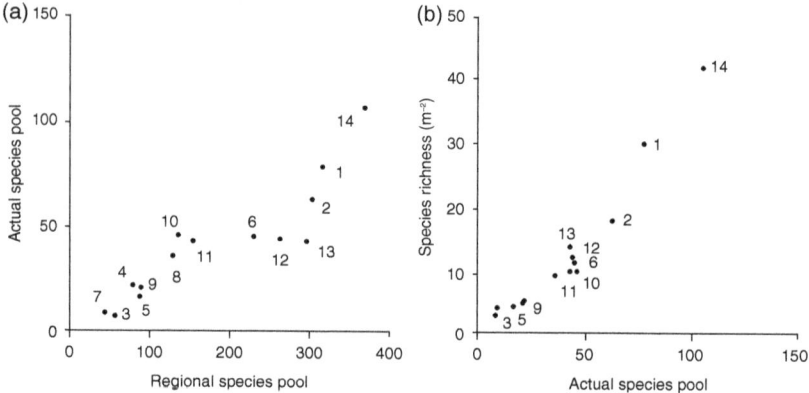

FIGURE 3.3 Data from Estonian plant communities shows that the number of species in a community increases with the size of the regional species pool (a). At the smaller scale, the number of species in a quadrat increases with the number of species observed in the community (b). (After Pärtel et al. 1996.)

Using amplitudes of 1.5 units, they obtained a four-dimensional hypervolume with a maximum distance from the centre of the hyper-volume being three units. The pool was constructed by "including from the regional flora (1) all indifferent species and (2) all other species for which the indicator values for light, soil moisture, pH and nitrogen content were within the interval of +/– three units around the average of that community in the four-dimensional hyper-space" (p. 113).

This study illustrates work arising out of a well-studied situation. The flora of Estonia is known. There is a long history of studying vegetation in Central Europe, with a rich array of other published studies. There is a compendium of data on the species pool, with key ecological requirements known for each species. The individual sites are small enough that they can be exhaustively surveyed. In most locations, we do not have such complete and reliable data.

We started off this section with the intention of reviewing the use of words in describing pools, hoping to provide a summary and some sort of standard terminology. We soon realized that this was an impossible task for a book of finite length and other topic areas. We don't want to get mired in definitions of the different kinds of pools that might exist. Nor do we want to offer an exhaustive list of names, with a new name for each kind of technique being used. Still, some consistency in nomenclature is necessary.

We suggest that the word "**pool**," without a modifier, be used in a general sense. At the largest biogeographic scales, we can modify the term and refer to the global pool or continental pools.

The term "**regional pool**" should be used mostly to describe a geographically limited area. Examples might include the peninsula of Nova Scotia, the Iberian Peninsula or the Gulf Coastal Plain of the USA. The global map of 867 ecoregions nested within 14 biomes (Olson et al. 2001) is likely to prove helpful for standardizing the geographical areas in defining regional pools. Recall that a regional pool is unfiltered – that is, it includes all species that could potentially disperse into a habitat even if they would be filtered out. These regional pools can be created by using published lists or by accumulating samples of communities across the region.

The term "**habitat-specific pool**" should be used to describe pools constructed solely for a single habitat. This use of the word pool is quite different, because it is constructed based upon data where the habitat filters have already acted to control species composition. That is, it is ecologically as well as geographically constrained. An example of this would be the species found in a particular kind of wetland, such as a wet prairie, a fen or a peat bog. Mostly such pools are produced by accumulating samples of the same habitat across a large region.

The term "**potential pool**" can be used to describe a pool that contains not only the current documented species, but also those that have the potential to invade the pool from other regions. This term may become more important as ecologists grapple with the challenges of invasive species, which we will explore at the end of this chapter.

We suggest that these terms will mostly suffice for describing levels stretching from the simple sample unit to the biogeographic realm. Our usage is not entirely consistent with work published to date. For example, Pärtel et al. (1996) used the term regional pool, whereas we prefer the term habitat-specific pool to make it clear that this type of pool includes species that are all adapted to the same habitat type. Pärtel et al. (1996) compared the observed actual pool to the habitat-specific pool, whereas we prefer to conceptually distinguish the community vector **C** from the species pool **P**: the local observed community is a subset of the unfiltered regional pool. There is one general lesson here: it is important to specify which kind of pool we are using in a study, what standard reference source has been used, and whether any specific rules have been used to expand or reduce the list provided in that standard reference.

SOME THEORY ABOUT POOLS AND COMMUNITIES

We began with large sources of published information on species pools, such as regional plant or bird lists. Now let us go to a much smaller scale and think a little more about how a species pool relates to the composition of particular habitats (or even to specified sample units within those habitats). After all, most of us who do field work have the personal experience of enumerating species in particular sample units. Eriksson (1993) provides a useful start. He explores the relationship between the number of species in a quadrat, or sample unit, and the number in the pool. He begins by asking about the relationship between S, the number of species in "any arbitrary unit space on Earth," and Q, the number of species in the pool. Our community vector **C** describing the biota of this arbitrary space on Earth is a set of species p_1 to p_S (recall Figure 1.1). This vector could be the contents of a pitfall trap, the list of plants in an alvar community, the list of birds on an island or, to return to the examples of Chapter 1, the mammals in one "landscape" in Kruger National Park, or the plants comprising an Ecoelement in the Ontario land classification system. For simplicity, we shall use the general term community. We use the word community in the sense that it is rather larger than a single quadrat, but not as large as Kruger National Park.

Erikson invites us to remember that the number of species in a community is a balance between two other factors, the rate of input of new species (c, for colonization) and the rate of loss of existing species (e, local extinction). Some of this thinking will be familiar to those of you who have read *The Theory of Island Biogeography*, which asks, in an analogous way, how many species will occur on a particular island, given an adjacent mainland and its species pool. Thinking now about any community, Erikson observes that the number of species is determined by the following equation:

$$dS/dt = c(Q - S) - eS. \tag{3.1}$$

That is, the colonization of species is proportional to the difference between the number of species already present (S), and those still available to arrive from the pool (Q). The extinction of species is proportional to the number of species already present (S), with some finite rate of local extinction. Setting the rate of change equal to zero, the equilibrium value of S, or the number of species in our community, is then

$$S^* = Q\left(\frac{c}{c+e}\right). \tag{3.2}$$

Erikson describes how this simple expression yields some predictions.

1. If the local extinction rate is very low, then S^* will be close to Q. That is, nearly all the species in the pool will be found in a particular community or quadrat.
2. If the local extinction rate nearly balances the rate of colonization of new species (that is, c = e), then the community will likely have about half the number of species in the pool, that is $S^* = \frac{1}{2}Q$.
3. If the local extinction rate is higher, with e > c, then the community will have even a smaller proportion of the species found in the pool.

These equations give us a way of thinking about just how many species in a pool are likely to occur in a specific location. With relatively long-lived species, like clonal plants, rates of extinction may generally be rather low, in which case the community may have a large proportion of the pool.

Each pool has many communities. And each community, in turn, has many representative quadrats. We can think of each small sample unit similarly as being in some sort of dynamic equilibrium with the community (and the pool). In this case, one could use similar equations, the first ones describing the relationship between the pool and the community, and the second that between the community and the quadrat. The equations will be similar, just nested into two levels.

A practical example can illustrate this. On Paul's property in Canada (protected by the local land trust) there are at least 10 small wetlands. Even though these wetlands are all within the same square mile of forest, and even though some are connected directly by streams, each has a visually different species composition. Let us think about each small wetland as a single sample unit, with the number of species being S. The above equations remind us that the species in each wetland change with time, with species arriving and species disappearing. The arrivals might be new plant species arriving in the guts of ducks, the disappearances might be the result of intense grazing by Muskrats. So we can think about each wetland as being like a small island. At the same time, however, the entire property has a combined set of wetland species that are themselves only a subset of all the wetland species found in Lanark County. A species can disappear from one pond without necessarily changing Q for the property. In the same way, a species can arrive new in a particular pond, either from an adjoining pond or from outside the property, but if that species is already present elsewhere in the property, Q again remains unchanged. Thus, any ecologist thinking about habitats in a landscape is dealing with a nested set of values of S, with each one sampling from pools at larger scales. There are two natural limits to this view of nature. At the largest scale lies **P**, the list of all the species that comprise the pool. At the smallest scale lies not just a single wetland community, but a single quadrat within that wetland. Each quadrat, and each wetland, and each reserve, has a rate of turnover, arising from inputs and outputs of species, and its own value of S. For most purposes, we select two scales out of this continuum, and focus on the relationship between the species pool **P** and the community **C**, as in Figure 1.1.

RATES OF ARRIVAL AND DEPARTURE IN THE COMMUNITY

The number of species in any community is only a subset of the pool. The lower number of species in the community is the result of two main causes. Some species may not have arrived from the pool yet – we call this an immigration effect, or by another name, dispersal limitation. Other species have arrived and then disappeared, an extinction effect. Hence, the subset of species in the community is a result of equilibrium between these two processes. This is just classic island biogeography. One important assumption in the above models is this: the rate of extinction increases with the number of species present. This is a statistical given, in one sense. The more species there are, the more likely one of them is to disappear. This, too, is a key assumption discussed in *The Theory of Island Biogeography*. This process is not, by the way, directly explained by a mechanism such as interspecific competition or disease – not at all: it is merely a statistical observation. Like the more pencils you have in your field bag, the more likely it is that one of them will be lost. It explicitly does not include the effects of biology.

There are also real biological reasons why such a relationship might occur. For example, as the number of species in a community increases, rates of extinction may also increase from competition or from disease. That is an added factor. Similarly, the rate of extinction may also depend upon external factors such as the frequency or intensity of periods of fire, drought or hypoxia. And the traits that different species possess will determine their vulnerability to these factors. So we need to be cautious in assuming that the relationships between local extinction rates and S is simply a question of the number of species. The equation is a starting point that deliberately excludes most biology. Those of us who study plant and animal interactions are documenting possible causes for local extinctions.

Here, however, we wish to look at the other side of the process: immigration. That is, inputs from the species pool into a community. We begin with the neutral model, that is, that the arrival of new propagules will indeed be simply a function of $(Q - S)$, where Q is the number of species in the pool and S is the number in the community (recall Figure 1.1). This is the straight line showing that the number of new arrivals falls with the number of species present (Figure 3.4).

But now we will add some biology. Consider traits for dispersal: some species in the pool will possess the kinds of traits that allow for more efficient dispersal. Overall, it is probable that species with traits evolved for enhanced dispersal will be the first to arrive, and that they will arrive rather quickly, shifting the line into a curvilinear form (Figure 3.4). This will tend to generally inflate values of S in the short term, and affect the kind of species found in **C**. With knowledge of traits, we know which species those are likely to be. And, simultaneously, individuals without long-distance dispersal mechanisms will

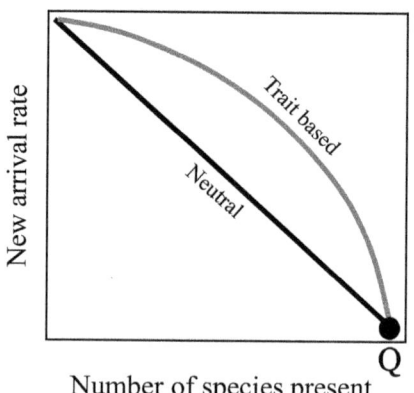

FIGURE 3.4 The neutral model assumes a straight-line relationship between the arrival of new species and the number already present. The presence of species with traits for dispersal will shift the immigration curve in a predictable manner. The actual number of species found in a community (S) will normally be much smaller than Q.

lag behind. Plants dispersed by birds, for example, are likely to arrive early, while plants dispersed by ants, less so.

When we think about this process, we can use the number of species present on the horizontal axis, with S being the total richness of the community. This is not the only possibility for the horizontal axis. We could also put Q at the end of this axis, extending it to include all the species in the pool. Or, we could replace the number of species by time. This might be useful when we are thinking about communities that are newly forming after a severe fire, or in an abandoned field, or on a new volcanic island. Each of these options gives a slightly different take on the events. We will stick with number of species present, mostly.

Now let us add in some biological interactions. We will also relabel the vertical axis to be explicit that we are really interested in actual establishment rather than mere arrival (Figure 3.5). This sleight of hand is justifiable, since the observed presence of a species in a community requires not just the arrival of individual propagules, but the survival and population growth of new individuals from that propagule. Given the prodigious seed production of plants (or spore production of fungi, or larvae production by marine invertebrates), there may be many arrivals, but the arrival of a propagule really only counts if one of them establishes a new population.

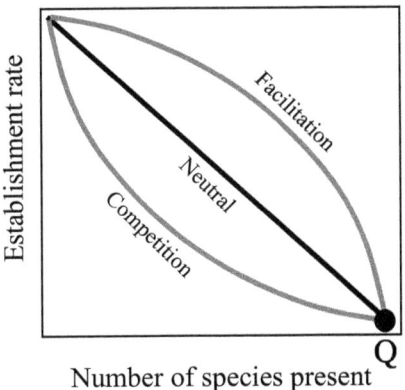

FIGURE 3.5 Facilitation will enhance establishment, while competition will reduce establishment.

In some cases, the presence of certain species on the site may actually enhance rates of establishment. This is known as facilitation. "Nurse plants" are a well-known example (Turner et al. 1966, Franco and Nobel 1989, Kellman and Kading 1992). So, facilitation may enhance survival and shift the curve higher. Soil mycorrhiza that are supported by existing plants may also be another source of facilitation (Wardle 2002, van der Heijden and Horton 2009, Keddy 2017). More generally, many kinds of mutualisms found in nature may increase local diversity in a community.

In most other cases, it is likely that the presence of species already in the community will be an obstacle to new establishment. This is generally the result of competition, which, of course, has a huge scientific literature (Diamond and Case 1986, Grace and Tilman 1990, Keddy 2001). Competition for resources will likely reduce rates of establishment, particularly in communities of sessile or territorial organisms where space is strictly limited (Yodzis 1978, 1986).

Biomass may also be a surrogate for effects of competition. There are well-described cases in plant communities in which the presence of accumulated biomass of vegetation itself feeds back to control S and C. The principle effect of that biomass is to produce a canopy of existing plants that reduce light for new arrivals (Grime 1973a, 1973b, 1979). Let us look at a situation where, instead of putting S on the horizontal axis, we use biomass of existing species (Figure 3.6). Here, the trait differences among species is also important – some plants have clonal growth and produce especially dense canopies and accumulations of leaf litter. These plants, when present, exclude many other species. In addition, early arrival by such species will give them an even greater local advantage. So, in this case, the probability of new species establishing is very much determined by biomass, and particularly the biomass of such clonal species. Hence, we may expect the rate of establishment to be particularly low on the right-hand side of the figure, since shade is often fatal to the establishment of seedlings (Harper 1977). We might therefore assume that establishment should be much higher at the low-biomass end of the gradient, where there is no canopy. But here it is possible that the low biomass is an indication that the site is unsuited to the growth of many potentially colonizing species. The competition from existing plants may indeed be low, but there is a reason for the lack of plants: lack of resources. So, in fact, the best conditions for establishment of new individuals may be at an intermediate position along this gradient, a region where neither physical stress nor competition predominate. This may partly explain why sites with intermediate levels of biomass have high diversity (Figure 3.6). That is, the corridor of diversity may reflect a situation where neither physical stress nor competition predominate. Grace (2001) has extended these ideas with a general exploratory model showing how biomass can regulate diversity in plant communities by controlling inputs from species pools. Fraser et al. (2015) have collected examples of this pattern at the global scale. Such work raises an interesting question: what might happen to the local community if we experimentally supplement rates of invasion by sowing new individuals from the pool into the community? We will look at some such experiments in the next section.

The main point here is that knowledge about the pool is a vital first step. And, general models exist that show us how immigration and local extinction can determine the likely number of species in a particular habitat, and the kinds of traits those species will possess. Once we have knowledge of species traits, and the environmental conditions, it is likely that real communities deviate significantly from neutral assumptions. In particular, species with high dispersal abilities are likely to be overrepresented. At least this sounds reasonable, until it is balanced by the fact that species able to dominate a site by clonal growth may eventually exclude these dispersal-adapted species. In which case, the value of S, and the kinds of species found in C, may be determined mostly by local conditions, particularly the rate at which competitive dominant species are removed to create gaps for new arrivals. This takes us into the realm of the role of natural

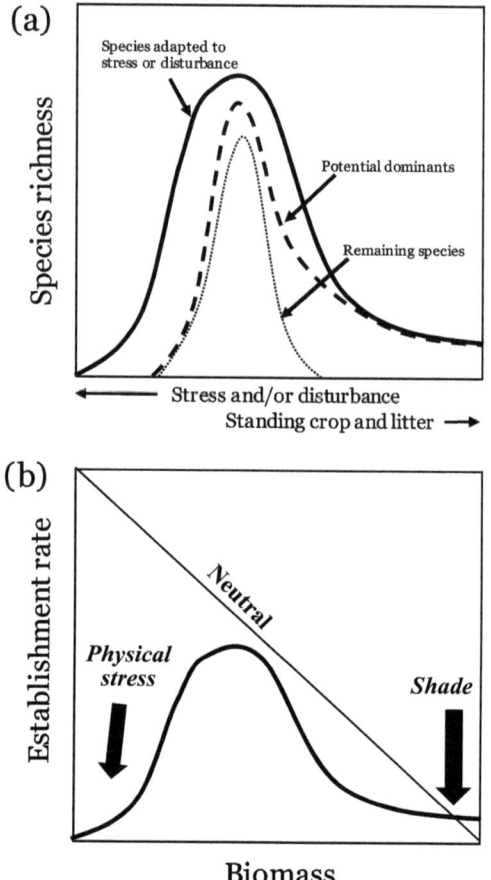

FIGURE 3.6 (a) Grime's (1979) intermediate biomass model suggests that in areas with low biomass, establishment may be low owing to physical constraints, while in areas with high biomass, establishment may be low owing to lack of light. (b) Hence, intermediate conditions will have higher rates of establishment and therefore higher S. ((a) adapted from Grime (1979).)

disturbances in generating local diversity. Natural disturbances including fire and grazing, as well as pulses of drought or flooding, will control the number of species in a community (Huston 1979, 1994, Pickett and White 1985). In the same way, natural disturbances will also influence the kinds of species (i.e., the kinds of traits) that are

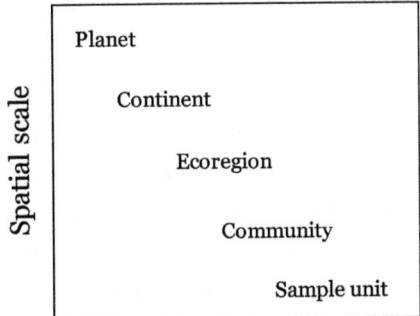

FIGURE 3.7 Species turnover in the community vector **C** is caused by local extinctions combined with the arrival of propagules. The rate of turnover will tend to be higher at smaller spatial scales. There are at least two underlying causes of this higher turnover. Small areas have smaller populations, and small areas are also likely to lie within the spatial scale at which natural disturbances typically occur.

found in the community. Many kinds of natural disturbance have a characteristic patch size, be they rather small (e.g., prairie dog mounds) or rather large (boreal forest fires). Smaller areas, particularly individual sample units, will tend to be smaller than the typical patch size, in which case major shifts in **C** (composition) and S (richness) will occur more frequently. Hence, in general, the smaller the spatial scale, the higher the species turnover rate is likely to be (Figure 3.7). We will soon look at actual experiments where immigration rates have been experimentally manipulated. But first, let us look at some general patterns found in species pools.

SOME EMPIRICAL RELATIONSHIPS OBSERVED IN REAL COMMUNITIES

We will now look at actual numbers for species pools in two relatively well-studied areas. The first example comes from an area that is frequently burned, the other from a habitat that is frequently flooded. Thus, we know that there are strong filters, and can focus on how the pool responds.

Example 1:   Species Pools for Plants in Mediterranean-
Climate Regions

Five areas of the world have Mediterranean types of climate with
warm, dry summers and cool, wet winters. In order of decreasing
size, these are the Mediterranean basin, California, southwestern
Australia (kwongan), central Chile and the cape of South Africa (fynbos). Table 3.1 shows the size of the species pool for each region.
Although these environments cover less than 5 percent of Earth's
surface, they have nearly 50,000 plant species, roughly 20 percent of
the world's total. Many of the species are sclerophyllous shrubs that
appear superficially similar. Cowling et al. (1996) hypothesize that the
high plant diversity is the result of two interacting causes: relatively
low growth rates and the reshuffling of competitive hierarchies after
fires (we will have more to say about competitive hierarchies in
Chapter 8). Herbivory is also an important filter in these habitats:
we have already seen in Figure 2.6 how grazing by goats as opposed
to elephants can produce different plant communities that arise out of
the same species pool.

There are some other distinctive features of the species pools in
Mediterranean areas. The southern hemisphere family Proteaceae is
very well represented in both the fynbos and kwongan. These two
areas also appear to have converged in a number of certain kinds of

Table 3.1 *Plant species pools in Mediterranean-climate*
regions (Cowling et al. 1996).

Region	Area (10^6 km^2)	Native flora (no. species)
California	0.32	4,300
Central Chile	0.14	2,400
Mediterranean basin	2.30	25,000
Cape	0.09	8,550
SW Australia	0.31	8,000

traits: a high incidence of species with obligate dependence upon fire for reseeding, serotinous seed storage in the canopy and seed dispersal by ants (myrmecochory). Certain genera have rapidly diversified into this habitat, including *Eucalyptus* (>300 species), *Acacia* (>400 species) and *Erica* (>500 species). This illustrates how filters can produce a particular set of traits, and yet large numbers of species with those traits can coexist.

According to (Cowling et al. 1996, p. 363), "the southwestern zone of the Cape has the highest diversity at this scale: for a given area, this region has, on average, 1.7 times the diversity of southwestern Australia, about 2.2 times the diversity of the southeastern Cape, California and the Mediterranean Basin, and 3.3 times the diversity of Chile" (Figure 3.8).

Example 2: *Species Pool for Plants in Intertidal Environments*

Although plant life originated in the ocean, plants invaded the land in the Silurian period some 400 million years ago. The first plants were small, without leaves, seeds or flowers, but all of these traits evolved

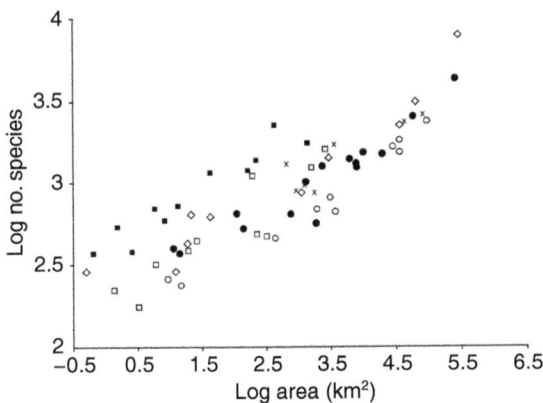

FIGURE 3.8 Species–area relationships in pools from Mediterranean-climate regions. ◊ SW Australia, ■ cape (SW) (South Africa), □ cape (SE) (South Africa), x Mediterranean basin, ○ central Chile, ● coastal California (After Cowling et al. 1996).

as natural selection from terrestrial environments acted upon these early terrestrial plants. Now the vast majority of terrestrial plants have leaves, seeds and flowers. It is noteworthy that the reverse process has been slow – very few terrestrial plant species have successfully reinvaded aquatic habitats, either freshwater or marine. In his monograph on aquatic plants, Sculthorpe (1967) observes that all present-day aquatic plants evolved from terrestrial ancestors, and have secondarily reinvaded rather than originated in aquatic environments. He estimates that less than 1 percent of the world's flora can be considered aquatic. Part of the explanation for this may lie in the constraints that aquatic habitats place upon plants, including anaerobic rooting environments, mechanical stresses associated with waves and low concentrations of dissolved carbon dioxide in water.

Saline water is even more of a barrier. Only a small fraction of the world's wetland flora, probably less than a further 1 percent, can occur in habitats that are simultaneously wet and saline. Salinity interferes with water and nutrient uptake, and most saline environments also have significant tidal activity, so that plants are cyclically exposed to an atmosphere and then flooded with saline water. Intertidal environments appear to have posed a nearly insurmountable obstacle to plant adaptation. For some numbers, let us focus on woody plants. A small number of tree species tolerate marine conditions and form intertidal forests (Tomlinson 1986); the usual nomenclature here is to call these *mangroves*, although Tomlinson called the vegetation itself *mangal*. Tomlinson reports that there are only 9 main genera of mangroves and some 34 species in the world, with 11 minor genera contributing a further 20 species. That makes a global species of 34 if narrowly defined, or 54 if we use a slightly broader definition. If we consider associates of mangroves, this list can be extended by a further 60 species, for a maximum of about 100 species. To put these figures into perspective, 90 woody plant species have been found in a single 500 m^2 transect in a tropical riparian floodplain of Central America (Meave et al. 1991). That is, just one transect in one freshwater floodplain had the same number of species as the world mangrove flora!

The constraints imposed by salinity are further illustrated by comparisons within plant families; there are some 200 genera and 2,600 species in the palm family, but only four species are commonly found in mangrove forests (pp. 30, 295). The family Myrsinaceae has over 1,000 species in about 30 genera distributed throughout the tropics and subtropics, but only four species occur in mangrove forests (p. 284). Similar sorts of calculations could be done with herbaceous families.

The combined filters of flooding and salinity appear to be very effective in reducing pool size. This is a reminder of our second proposition in the preceding chapter: in any habitat, only a small number of filters is likely to be important. Indeed, continuing with the example of cold, if you add in a third factor, freezing resistance, there apparently has been no evolutionary solution for woody plants. Flooding with salt water, combined with occasional freezing weather, removes all woody plants from intertidal habitats. Salt marshes, dominated by herbaceous plants, mostly grasses, replace mangroves when these three filters are combined. You can visit marshes in southern Louisiana that are right at the transition point: the periodicity of cold events is just enough to kill mangroves back to the soil line, but not enough to kill them outright. Hence, at that latitude, one can stand in the very location that divides extensive salt marshes to the north and extensive mangrove swamps to the south. The same pattern is evident in Florida, where only the southern tip of Florida has mangrove swamp.

Now, here is an evolutionary question. To start with, why is it that so few plants have evolved tolerance to flooding and salinity? It is not possible to say that the ability to tolerate both flooding and salinity is a problem without an evolutionary solution, since in a small subset of species, natural selection has clearly produced physiological solutions to the dual stresses imposed by the intertidal habitats. Part of the answer may lie in energetics (e.g., Hall et al. 1992), that the metabolic costs of tolerating either flooding or salinity alone are high enough that it is difficult for species to maintain a positive energy balance, particularly when the usual costs of foraging for resources, competing with neighbours and defending against herbivores must be simultaneously addressed. When these costs are all added together, the added metabolic costs of flooding and salinity may be sufficiently high that only a small subset of species can forage and photosynthesize rapidly enough to successfully balance these expenditures. And, apparently, the addition of one added filter, frost, makes an evolutionary solution impossible. Hence, the species pool for mangroves in Nova Scotia and Scotland is exactly zero.

There is one interesting difference between our two examples. Although both habitats have produced one or more functional types adapted to deal with the local environment, Mediterranean vegetation has large numbers of species, while mangrove vegetation has a relatively small number of species. Why this should remain an open question? It does point out a recurring issue in community ecology: although environmental filters can produce a small number of traits and/or functional types in a landscape, some other factor seems to be determining how much speciation will occur within each functional type. This theme of diversity among and within functional types clearly deserves more attention, since it has a big impact upon species pools.

EXPERIMENTAL MANIPULATION OF IMMIGRATION AND ESTABLISHMENT FROM POOLS

We now have a growing body of experimental data (from more than 60 studies!) that allow us to better understand how communities arise from pools (Myers and Harms 2009). The general conclusion is that the flow of propagules from the pool into a community has measurable effects on the diversity and composition of the community. The standard procedure is to increase the seed supply into a community, and then measure the consequences. A meta-analysis of 62 such experiments across grasslands, forests, savannas and wetlands clearly demonstrated that seed arrival increases local species richness: 70 percent of experiments observed a significant increase in species richness after seeding compared to control plots. These results make an important point: many communities are not fully saturated. That is, dispersal limitation limits the diversity of the community. In other words, there is often room for more species, but they just haven't arrived yet.

Most of the studies in the meta-analysis occurred in northern hemisphere temperate ecosystems, but similar results were found in a tropical rain forest as well. Paine and Harms (2009) added seeds of canopy species to a moist floodplain forest in Peru's magnificent Manu National Park. They found that sowing density was the best predictor of seedling density and that seedling richness was mostly determined by the seed addition. This study also found that seed supply affected the composition of the seedling community, a reminder that just a single seed and single arrival event could trigger a shift the composition of the seedling layer which could profoundly impact the future composition of the forest canopy.

Another important discovery was that propagule supply interacted with environmental filters (Myers and Harms 2009). We speculated above that competition could impact rates of immigration by controlling which species successfully establish. Experimental disturbances do indeed create opportunities for establishment, presumably by temporarily reducing competition. Disturbance enhanced the positive effect of seed supply on species richness by 70 percent.

Consider the effects of fire as a recurring natural disturbance. The longleaf pine savannas of the southeastern USA and Central America are similar in some ways to Mediterranean plant communities in Africa and Australia, having large numbers of species in communities, and also relatively large species pools. These pine savannas are also of high conservation concern (Williams 1989, White et al. 1998). How best to restore plant diversity in degraded sites? Recreating natural fire cycles is a well-recognized procedure (Glitzenstein et al. 2003). We now have an example of both fire and seed inputs being simultaneously manipulated (Myers and Harms 2011). High-intensity fire temporarily reduced species richness in sites that did not receive additional seed, but high-intensity fire had virtually no effect on species richness in sites that received the seed additions. This raises the question about the degree to which limited propagule dispersal from the larger pool may be limiting habitat restoration, particularly when many small sites that are being targeted for restoration are isolated from other natural areas. It may be the combination of a natural disturbance and the addition of propagules will be more effective in restoring natural levels of diversity to degraded ecosystems.

INVASIVE SPECIES AND THE POTENTIAL SPECIES POOL

The problem of invasive species is closely related to the topic of pools, and to changes resulting from anthropogenic activities. The biogeographic realms and world floristic regions were once relatively isolated from one another. Now pools are being augmented by new species from other realms. Let us return to the example of Lanark County in Ontario. A European orchid, *Epipactus helleborine*, is now common in the forest understory. A European shrub, *Rhamnus cathartica*, is invading along old settlement roads and then being locally dispersed by birds. Beech bark disease, triggered by the European Beech Scale (*Cryptococcus fagisuga*), is now killing beech trees and thereby changing light conditions for other trees and plants. An Asian beetle, the Emerald Ash Borer (*Agrilus planipennis*) has begun killing ash trees. All future communities will be changed as a result of these invasions.

Humans are now one of the most effective means for long-distance transport, and the rapid establishment of new species in landscapes is a phenomenon that cannot be overlooked in the issue of community assembly. There is, increasingly, a great risk that species will indeed escape from certain biogeographic barriers and enter new species pools. The addition of new and invasive species to pools may be one of the most dangerous and long-term effects of humans upon ecological communities. Consider the way this biological pollution differs from industrial pollution. Most industrial pollution will, over time, decline when inputs of the pollutant stop. Industrial pollutants decline at a rate determined by exponential decay. Invasive species, on the other hand, once introduced, do not decline, but expand at an exponential rate. The problem gets worse with time.

Ideally, our predictions of potential future composition in communities should in some way address the issue that most habitats have a list of potential colonists merely waiting for the right opportunity to become part of the species pool. It can happen very quickly. In only a few decades, the wetlands in Lanark County have become dominated by an invasive floating aquatic plant. The European Frog-Bit (*Hydrocharis morsus-ranae*) was historically not a part of the species pool for wetlands in eastern North America. However, it escaped into the Rideau River from the Central Experimental Farm in Ottawa around 1939 (Catling and Dore 1982) and has now become a well-established species in wetlands around the Great Lakes. Beaver ponds that were once mostly open water are now covered in a floating canopy. This is just one of countless examples, with more prominent global examples including goats introduced on the Galapagos Islands, starlings introduced to North America and pythons introduced in the Florida Everglades.

If community ecology is to be able to say something predictive, rather than reacting individually on a case-by-case basis, there needs to be some method for delineating species that are not currently in the pool, but which have the appropriate traits to invade and become part of existing communities. Hence, the importance of recognizing the existence of a *potential pool* for any particular set of habitats. At one extreme, the potential pool might be as simple as a set of native species with distributional limits that do not yet quite extend to the habitat of interest. At the other extreme, it might include species that are currently far away, but which possess the traits to rapidly establish and spread in the case of a single dispersal incident.

APPROACHES TO MEASURING POOLS

As noted at the beginning of this chapter, we have, in many cases, been provided with lists of species that provide the species pool for a wide array of species and locations. We therefore have abundant data to work with. At the same time, the data are fuzzy, because it is not always clear which species are actually part of the pool. Some of this is the result of natural historical complexity, including details of dispersal. Some of the fuzziness arises out of issues of scale, and the meaning of words. Let us look more carefully at the process of collecting data on pools. We will begin with the classic case of islands with an adjoining mainland since it should be the most obvious situation, and therefore best able to illustrate how ambiguities arise. Then we will discuss two general approaches to delineating species pools: species accumulation and downscaling.

Islands and Mainlands

Consider a series of islands with an adjoining mainland. In this classic situation, the list of species from the mainland, more generally called the "source region" in MacArthur and Wilson (1967, e.g., figures 12–15), is usually assumed to be the species pool. This is the approach used in many applications of *The Theory of Island Biogeography*. A particular example is Diamond's (1975) enumeration of island bird faunas as subsets of the source region pool, the avifauna of New Guinea.

Even in these apparently clear-cut cases, with a well-known mainland pool, there is nuance. Some species found on the mainland list are probably unable to live on islands owing to habitat characteristics of those islands. Should they be counted as part of the pool? Other species may be unable to disperse there. Should they be counted as part of the pool? Consider birds, and the peninsula of Nova Scotia on the northeast edge of North America. An example of the first situation would be forest interior birds, such as the Wood Thrush, which occur in forested areas of south-central Nova Scotia, but probably cannot find sufficiently large tracts of forest in small islands. Indeed, this bird does not even breed on the largest island, Cape Breton Island. Large woodpeckers would be another example of a species unable to feed or nest on offshore islands lacking large trees. An example of the second situation would be flightless birds. The Ruffed Grouse and Spruce Grouse are not entirely flightless, but nearly so.

These exemplify two factors that complicate the definition of pools: the tolerance limits of the species and their dispersal abilities. Hence, the real pool of candidate species that might be found on islands is probably smaller than the full list of species found on the mainland. How much smaller? That is the interesting question, one that generally requires careful consideration. Note that in both cases, whether one is considering tolerance limits or dispersal abilities, we are actually talking about the traits of each species. Once we have removed species that cannot survive on islands, or cannot disperse to islands, we have a rather smaller pool.

Birds that are restricted to forest interiors, and birds that cannot fly, may be a minor problem in assessing the mainland pool in birds. The problems remain, and likely become even more difficult, when we move to consider species other than birds. Which mainland plants are able to actually survive on islands? Many are small and windswept with shallow soil, which likely excludes many common deciduous tree species – and therefore species that live in association with them. Many offshore islands, if forested, have only a few species of conifers. Which mainland plants are able to disperse to islands? In most cases it is difficult to say, and requires us to measure multiple possible methods of dispersal, including being carried by wind, ferried in the guts of birds or possibly drifting in the ocean. Early naturalists paid a lot of attention to the problem of dispersal. We quoted earlier from observations on animals transported in floating islands of debris. Charles Darwin actually conducted experiments on the ability of plants to tolerate immersion in sea water for extended periods of time (e.g., Darwin 1855). It therefore seems to be the case that without systematic screening of traits for multiple kinds of dispersal, we cannot say in advance which species are or are not candidates for dispersal to or survival on islands.

A further problem is that, in most cases, the answers to questions about tolerance limits and dispersal do not give clear yes/no answers, but rather probabilities. Hence, clear-cut rules may be difficult to find. At best, we may be able to rank species in terms of relative ability to tolerate island conditions, or relative abilities to disperse to islands. If we are working with plants (or beetles), we are dealing with an order of magnitude more species. So, in general, it is necessary to face up to the difficulty of defining the pool. It is smaller than the mainland pool, but how much smaller? What kinds of measurements must we make to answer the question?

Hence, in practice, in the short term, the most practical option is not to start with the mainland, but rather to visit all the islands and compile the list of species that actually occur on islands.

Even this approach has its hidden assumptions. During ice ages, sea levels are much lower, and extensive areas of the continental shelf become land. Even isolated Sable Island in the Atlantic Ocean, now little more than a series of dunes, with no native tree species, has formed on what was once continental shelf. So many islands were once part of the mainland, and have been isolated from the mainland by rising sea levels over the past 10,000 years. Hence, plants that currently are found on islands may not have reached there by long-distance dispersal, but rather are remnants from a time when the island was not an island at all.

At this point, it is probably worth reminding ourselves why we are on this topic. It is not just that we are interested in offshore islands. We are exploring how we might construct verifiable and accurate species pools. We have picked islands and mainlands because they are a limiting case where the task might seem to be relatively simple. If the task is difficult for this particular case, then it might be expected to be far more difficult for patches of different habitat scattered across a continent.

As if these problems are not sufficiently perplexing, there is one further issue. In some cases, there may also be exceptional cases in the other direction – that is, species that occur on islands but not on the mainland. Let us consider three natural history examples.

In Nova Scotia, for example, the endangered Mountain Avens (*Geum peckii*) was for many years only known from one island, Brier Island, where it grew in unforested peatlands (Roland and Smith 1969). Although not known from the mainland, it was certainly part of the species pool for islands. But then, from another point of view, Mountain Avens really is part of the species pool on the mainland, if you simply make the mainland big enough. The other location for this plant is relatively far from Nova Scotia, in high-altitude peatlands in the White Mountains of New Hampshire. So one could argue that Mountain Avens would indeed be part of the mainland pool if the mainland were simply made geographically more extensive to include the mountains of New England. This begs the question of just how big an area the mainland needs to be. And now such questions are simplified because one small population has recently been found on the mainland of Nova Scotia itself. Many years of extensive field work by botanists had not located any other populations of this plant, either on other islands, or in similar habitats on the mainland. Then a small population was located on the mainland, and, although its long-term survival remains in question, its status in the species pool has changed again (COSEWIC 2010).

Again in Nova Scotia, the Ipswich Sparrow breeds only on Sable Island. Hence, it does not occur in the mainland pool. But it would occur in the pool for islands if we tallied up all the bird species that currently live on islands. It is also observed on the mainland during migration. And, more nuance, Ipswich Sparrows are now regarded as only a subspecies of the more widespread Savannah Sparrow (Wheelwright and Rising 2008), in which case the species is represented in the mainland pool. Sort of.

A third example, from the other side of the world, is the Tuatara Lizard, endemic to New Zealand. At one time, the species was common on the two main islands, called, appropriately enough, the North Island and South Island. Like the flightless moa, Tuatara were exterminated from the mainland islands by the indigenous Māori. Unlike moa, Tuatara survived on offshore islands. The principle behind this pattern is the same as we saw with amphibians in ponds in Chapter 2 – the importance of habitat free from predators. Islands provided such space. If conservation efforts to eliminate predators from the mainland succeed, they may again become part of the mainland pool. Indeed, populations now survive on the mainland in areas that are fenced and predator-free.

Although these three cases may seem to be just stories of natural history, they illustrate how difficult it can really be to decide what the species pool contains, particularly in larger groups such as plants and insects. To date, each investigator often chooses a particular solution to their particular situation, which can lead to lack of consistency in methods, terminology or both. Our general recommendation is to choose a standard source, like the examples given in Figure 3.1, so multiple researchers are at least using the same source. One could then modify it, where needed (more on this below). And, one can keep in mind that, like all ecological measurements, there is a margin of error.

Island situations with mainlands are attractive because they illustrate a limiting case – highly discrete patches surrounded by inhospitable conditions. To achieve more generality, we must consider the problems with constructing a pool for situations where our habitat may indeed consist of patches scattered over the landscape, but where adjoining habitats are a potential source of colonists, and where there is no clearly defined mainland pool. How might we proceed?

The Accumulative Approach to Creating a Species Pool

We could try to estimate the pool for a habitat by accumulating data on the number of species in samples or patches; since we are building the pool from small samples, we can call this the accumulative approach. One of the most ubiquitous measures in ecology is the number of species in a sample unit, whether that unit is a quadrat, benthic sample or transect. The number of species in a small quadrat

APPROACHES TO MEASURING POOLS III

is known as alpha diversity or species richness or species density (e.g., Whittaker 1965, Pielou 1975). We know generally that there is a well-established relationship between the area of habitat we sample and the number of species observed. One example is the species–area relationship, such as the one already shown for Mediterranean plants in Figure 3.8, where each point on the line represents a different island or area of habitat. When we talk about species accumulation curves (also called species discovery curves), we are talking about something related, but slightly different. We are interested instead in the cumulative number of species we find as we accumulate more and more samples from the same habitat. Each sample unit may be, and often is, the same size, but we are accumulating more of them. So, this is, more generally, the relationship between number of species encountered as a function of sample effort. Effort can be measured in different ways, such as the number of quadrats enumerated, the number of hours of observation or even the number of individuals counted.

Consider starting within a single patch of habitat. When we plot the cumulative number of species we observe against sampling effort, we expect to find that the number of species we observe rises asymptotically to Q (Figure 3.9). At this asymptote, we should have a list of the species in the pool or, more precisely, the realized list of species in the pool. In practice, the asymptote takes a great deal of sampling effort. Consider the case of spiders in the Appalachians, in two specific habitats: heath balds, which are shrub-covered, and grass balds, which are covered mostly by herbaceous species (Toti et al. 2000). In each habitat, at a single location, 80 hours of effort were insufficient to find the total number of species (Figure 3.10)! For many years, people have tried to think of statistical procedures that use the shape and composition of the accumulation curve to try to guess the true number of species in the habitat (e.g., Colwell and Coddington 1994, Witman et al. 2004). We will not explore this here, since it is a large topic, and in any case, we are looking at the challenge of compiling a list of species in the pool. That is, even if there were a perfect tool for accurate estimation of Q, the number of

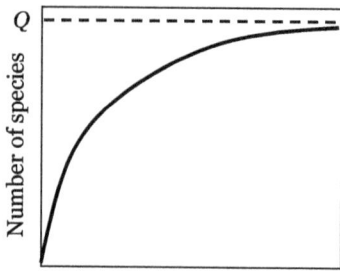

Number of sample units

FIGURE 3.9 The species accumulation curve illustrates the challenge of compiling a list of species in a habitat from increasingly larger numbers of sample units. The horizontal axis may be the number of quadrats or traps, or a surrogate such as number of hours of observation, or even the number of individuals identified.

species in the pool, it would still not provide the *list* of all species comprising this number.

How much sampling is enough? Consider a completely different habitat, the epifauna of vertical rocky surfaces on marine shorelines. Witman et al. (2004) found asymptotes in northern habitats such as the Gulf of Maine (sample sizes of 30 quadrats yielding about 20 species), but in tropical waters 30 quadrats were insufficient. In the Palau Islands, there were >300 species and no asymptote was found after 70 quadrats. Spiders in the Appalachians and invertebrates in the Palau Islands are just two of many possible examples that illustrate the challenge of finding the pool from accumulating many small samples. It requires a great deal of effort, with no guarantee of a full list. Indeed, even Witman et al. (2004) needed identifications, and had to compile a standard species pool "assembled from published species lists and by consulting taxonomic experts in each region" (p. 15666).

Downscaling as Another Approach

Given the challenges of working from the bottom-up by accumulating sample units, let us revisit the idea of mainlands. Let us return to the approach that uses a reference source, which might include data collected over decades and different habitats and by many researchers.

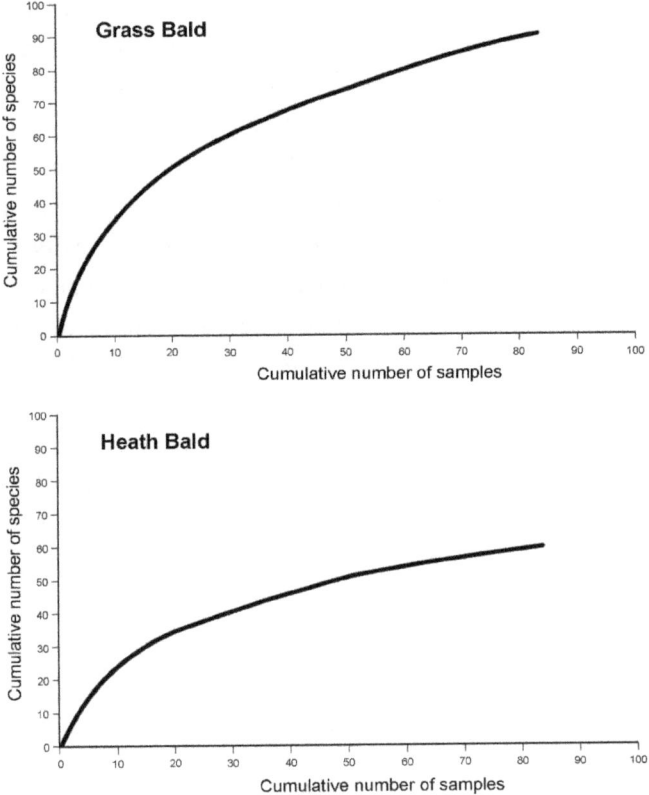

FIGURE 3.10 Species accumulation curves for spiders in two different Appalachian mountain habitats. In neither case is there an asymptote, a complete species pool for the location. (After Toti et al. 2000.)

We will then use their cumulative set of records for that region as the pool. That is, back to Figure 3.1, but with some added guidelines. The challenge is to reduce the geographic scale by looking for a list of species compiled for a smaller geographical area, such as a county or a specific park. We might call this the downscaling approach because we are still looking for a comprehensive and pre-existing list of candidate species, but one that is more narrowly confined to a particular region. Many areas of landscape now have such checklists. This is particularly true for rather popular groups of organisms in rather popular parks. The checklist of birds in Algonquin Provincial Park (Tozer 2012), for example, lists all 278 species of birds seen in the park, 144 of which are breeding species. The online list of mammals for Kruger National Park gives us 150 species, ranging in size from elephants to shrews. Of course, such lists include all the species in the region, not just species found in a particular type of habitat.

Even with some downscaling, the pool will usually be rather large. Returning to the tops of mountains in the Appalachians, our spider (or plant) species pool will be inflated because a checklist will include species that occur in the valleys but which could not survive at high altitudes. That is to say, the pool would include many species with fundamental niches that do not include ridge conditions. This is analogous to interior forest birds that occur in the pool, but which really could not be expected to survive on islands. But, in many cases, we cannot easily identify and remove such species from the list because it is unlikely that we will have information on the habitat requirements of large numbers of species at any time in the near future. Yes, the familiar problem of which birds could actually survive on offshore islands shows up again, be it with spiders on balds or beetles on alvars.

MORE ON OLDER LISTS AND THEIR UTILITY IN DOCUMENTING SPECIES POOLS

We began this chapter talking about the value of modern compilations that provide lists of species pools at large scales (Figure 3.1). We have just mentioned the further value of smaller compendia, such as the list of bird species in Algonquin Provincial Park, or mammal species in Kruger National Park. This seems the appropriate time to remind you that there are many, many other lists in the world of ecology. At one extreme, some are personal, like the lists of species in field notes from study areas. We might include in this category lists that are recreational, like personal "life lists" for birds. At the other extreme, some lists are extremely important, and are team efforts, like the list of species in the IUCN Red List, or official lists of species pools for ecological regions. The map of world floristic regions in the next chapter (Figure 4.2) is an example of how lists can generate global maps, while lists of species in different Mediterranean habitats allow us to explore large-scale geographical patterns (recall Figure 3.8). There would seem to be some need of discernment in deciding which lists are useful to science as a whole, and which are incidental.

Let us say more about the lists compiled by regional experts, particularly older lists. Often these are foundational for compiling accurate data on species pools for given regions or habitats. Table 3.2 gives one of many possible examples – the occurrence of amphibians in different regions of eastern Canada. Let us mention a few other related examples. Bleakney (1958) also provided a similar table for reptiles in these regions, while Cameron (1958) documents the distribution of mammals on the nearby islands of the Gulf of St. Lawrence. Some older papers have similar tables, while other older papers include lists of species within the text. For example, Raymond (1950) describes the vegetation of different regions of Quebec, with comments on the post-glacial migration routes and local habitats, and often provides lists of species considered characteristic of particular habitats like peat bogs, or particular watersheds, such as the Ottawa River valley and the Richelieu River valley. Two more recent examples were mentioned in the preceding chapter – the tabulation of plant species found in alvars and rock barrens in Ontario. Such lists provide us with expert opinion about the pools of different habitats within a region, what we call the *habitat-specific pool*. Some of these older studies may also provide raw material for calculating the power of various filters (Equation 2.1). The baseline information they contain may also prove useful in the future as species distributions change in response to changing climate.

The preceding examples are a reminder that knowledge of species pools requires a sound appreciation of natural history. Using the same ecological regions, the confirmation of a suspected native population of Blanding's Turtles in southwestern Nova Scotia (Bleakney 1958, Lefebvre et al. 2012), or the discovery of a new native species of

Table 3.2 The occurrence of amphibians in different ecological regions of eastern Canada (Bleakney 1958).

Amphibian	South-central Nova Scotia	Annapolis Valley, N.S.	Musquodoboit and Gay R. valleys	Oxford, N.S.	Cape Breton Island	Cape Breton Island Plateau	Prince Edward Island	South-central New Brunswick	Northern New Brunswick	Gaspé Peninsula, Quebec	Anticosti Island	Newfoundland	Eastern Ontario	Gatineau Valley	St. Lawrence Lowlands	Southeastern Quebec	Southern Laurentian Mts.	Saguenay Valley, Lake St. John	North shore of Gulf of St. Lawrence	Lake Melville, Labrador	Knob Lake, Menihek Lake	Ungava to the tree-line
Necturus maculosus maculosus	C	?	?	?	–	–	R	C	C	C	–	–	C	–	C	–	–	C	C	R	–	–
Ambystoma jeffersonianum	C	?	?	C	C	?	C	C	C	C	–	–	C	C	C	C	C	–	C	–	–	–
Ambystoma maculatum	C	C	C	C	C	A	A	C	C	C	–	–	C	C	C	C	?	–	–	–	–	–
Diemictylus v. viridescens	–	C	C	C	C	A	–	C	C	C	–	–	C	C	C	C	C	–	–	–	–	–
Desmognathus fuscus fuscus	C	–	C	C	–	C	R	C	–	–	–	–	–	–	–	A	–	R	R	?	–	–
Plethodon cinereus cinereus	R	R	C	C	R	C	R	C	C	C	–	C	C	C	C	C	C	R	–	–	–	–
Hemidactylium scutatum	R	R	–	–	R	–	R	–	–	–	–	–	C	R	–	–	–	R	–	–	–	–
Gyrinophilus p. porphyriticus	–	–	–	–	–	–	–	–	R	–	–	–	C	–	–	–	–	–	R	–	–	–
Eurycea b. bislineata	–	–	–	–	–	–	–	C	R	C	–	–	A	C	?	C	C	C	R	C	R	–
Bufo americanus	C	A	C	C	C	?	C	C	C	C	C	–	C	C	C	A	C	C	C	C	R	–
Hyla crucifer crucifer	C	C	C	C	C	?	C	C	C	C	C	–	C	C	C	C	C	C	R	?	?	–
Hyla versicolor versicolor	–	–	–	–	–	–	–	R	–	–	–	–	C	R	C	–	–	–	–	–	–	–
Pseudacris nigrita triseriata	–	–	–	–	–	–	–	–	–	–	–	–	C	R	C	R	C	–	–	–	–	–
Rana catesbeiana	A	C	?	C	–	–	–	C	–	C	–	C	C	C	C	R	C	?	R	–	–	–
Rana clamitans	C	C	C	C	C	C	A	C	C	C	C	–	C	C	C	C	C	C	C	–	R	–
Rana septentrionalis	C	C	C	C	C	?	?	C	C	C	C	C	C	C	C	C	C	C	C	R	R	R
Rana sylvatica	C	C	C	C	C	?	C	C	C	C	–	–	C	C	C	C	C	C	R	C	R	R
Rana pipiens pipiens	C	C	C	C	C	?	C	C	C	C	C	–	C	C	C	C	C	C	C	C	–	–
Rana palustris	A	C	C	C	C	?	?	C	C	C	C	–	C	R	?	C	R	–	–	–	–	–

wetland plant, Tall Beak-Rush (*Rhynochospora macrostachya*) in Nova Scotia (Blaney and Mazerolle 2009), or the discovery of Bear Oak (*Quercus ilicifolia*) in the rock barrens of eastern Ontario, all provide new information on species pools. These examples illustrate how continued field work is expanding knowledge of species pools even in areas of the world with more than a century of scientific exploration.

   We emphasize that each region of Earth likely has a similar mix of older scientific compendia supplemented by new discoveries. Therefore, the modern study of species pools requires familiarity with two quite different sources of information: the older biogeographic and descriptive literature (such as Bleakney (1958) or Cameron (1958) in eastern North America) as well as more recent reports of new discoveries. The older studies are easily overlooked, and may even not be available in electronic versions (and also may be missed in key word searches), while the newer studies are likely published in regional journals devoted to natural history and field ecology. Therefore, younger scientists working on species pools need to keep in mind the availability of these complementary sources of information. Although our examples come primarily from eastern Canada, we wish to emphasize that the same principles apply worldwide, particularly in areas that are poorly explored and subject to continued field inventory. In the Greater Mekong area of Southeast Asia, the World Wildlife Fund for Nature (2014) reported 267 new species, including 290 plants, 24 fish, 21 amphibians, 28 reptiles, 1 bird and 3 mammals!

   Anyone who has tried to personally compile lists for a specific region, or a habitat within a region, will realize the amount of work involved. We have made the case that older papers may provide important sources of reliable data. Of course, in other cases, one might find reports that are hopelessly inaccurate and generate hours of confusion. The Flora of Long Point is an example of the challenges. This huge sand spit on the north shore of Lake Erie is an important natural area with many rare plant and animal species. When Reznicek and Catling (1989) set out to document the flora and the ecological communities of this sand spit, they first had to survey the existing communities and document the new species they found. Then they had to painstakingly exclude more than 70 species that had been reported from Long Point, but for which there was no reliable evidence that they even occurred there (see their table 3)! Discernment is always necessary, otherwise past errors are transmitted through the scientific literature.

In a lamentable yet growing number of cases, these older lists written prior to the 1960s are irreplaceable for another reason – because the habitats described in these papers have now been ruined by human activities. The remarkable coastal plain communities of southwestern Nova Scotia, which we will revisit in the closing chapter, were documented a century ago by a team of botanists including M. L. Fernald (1921), and have since been damaged by several hydro-electric dams, highways, cottage subdivisions, mink farms, all-terrain vehicles and clear-cut logging. These changes since the Fernald expedition are just one specific case of a global phenomenon: the destruction of natural areas, which is reducing the species pools of many regions, and is driving the growing list of species on the IUCN Red List.

If we do not use this older literature, it is possible that we risk seriously underestimating the original biodiversity of native landscapes. Another example of knowledge from old expeditions: you can read (online) about the habitats and species William Bartram (1791) described during the 1770s while travelling on horseback some 2,400 miles across southeastern North America, beginning in South Carolina and ending in Louisiana. He was not travelling in cheerful times: these years were marred by the American War of Independence, and Bartram arrived in Louisiana while the British army further north was attempting to capture the Hudson River Valley, a venture that ended unhappily (for them) at the Battle of Saratoga on 17 October 1777. Bartram himself arrived in Louisiana severely ill, "incapable of making any observations, for my eyes could not bear the light, as the least ray admitted seemed as the piercing of a sword, and by the time I had arrived at Pearl river, the excruciating pain had rendered me almost frantic" (pp. 419–420).

Bartram was seeing species that have now disappeared from ecological communities there, including two birds, the Passenger Pigeon and the Carolina Parakeet. Let us have one more example of a missing species, a wetland plant. In 1912, an American botany professor, Norma Pfeiffer, discovered a species new to science, *Thismia americana*, in a wet sand prairie near Lake Calumet in Chicago (Pfeiffer 1914). She saw it for five consecutive years. It has never been seen again. Since then, the habitat has been swallowed by urban sprawl. Lake levels and climate may also have changed. Meanwhile, teams continue to search for it in remnant prairies (Bowles et al. 1994, Rodkin 1994).

Many more such stories of missing species could be told, and we invite you to find an example of a species that is now missing from the species pool in the ecological region where you live. Such species should surely be mentioned in our teaching, as local examples of a pressing global problem. Loss of biodiversity might appear to be an abstract concept, vaguely associated with foreign habitats, until students become aware that there are already species that are missing from their own landscape.

Owing to changes in style, lists and vivid habitat descriptions are less common in modern scientific papers. We are told that many younger scientists do not read older work. Yet, in some cases, such work provides important information about regional pools, particularly regional pools including species that are now rare or even absent from present communities. These older studies may also provide insightful hypotheses about causal factors that we now might put to more formal tests. The challenge of incorporating this older work into modern studies of species pools remains open.

In summary, within any particular region, ecologists have a rich array of sources, including modern compendia (Figure 3.1) and older papers from selected habitats (Table 3.2). These are uneasily nested in some sort of spatial hierarchy, and they are controlled by filters that may not yet be specified. They are important raw material for community ecology, so long as they represent competent field work.

Of course, not all lists matter. Discernment is always necessary. Tansley himself was insistent that ecologists should not be compiling and publishing random lists (1914, p. 200): "Thus if one goes into an oak-wood or on to a heath and makes a list of the species of the ground flora, that may be a necessary part of an ecological investigation, but to publish such a list, unless for some definite further purpose, is perfectly useless as a contribution to ecology." While we generally agree with Tansley, we would add that with some hindsight, competent lists for specified regions or habitats that are now buried in the older literature may indeed have a "definite further purpose" – as important raw material for constructing species pools. How many data sets like those in Table 3.2 lie buried in publications from the preceding century? How many new species remain to be documented in areas like the Greater Mekong?

LOOKING AHEAD

To conclude this chapter, we wish to draw attention to the urgent need for standard lists of species pools for different taxa and different regions of Earth. Although we have tried to emphasize the positive – the existence of standard manuals that list species for many regions (Figure 3.1) – we have also described the many limitations of these data. Lacking standard data for pools, ecologists will be forced to create them on a case-by-case basis. This can easily lead to lack of consistency in criteria, and to the proliferation of nomenclature for kinds of pools. We have noted that the habitat-specific pool appears in the literature under other names. On the positive side, the situation may be similar to that which existed for trait matrices only a few decades ago. When Paul first wrote about the importance of trait matrices for community ecology (Keddy 1992, 1993), such trait matrices were scarce. Now, as we will see in the next chapter, they are growing rapidly in size and availability. We anticipate a similar possibility for data sets on species pools.

One of the particular challenges for species pools is the tendency for much of the literature to use political rather than ecological boundaries. We have done so ourselves in this book, mostly because that is how the information is currently stored. However, we have also seen the inadequacies of this system – for example, Figure 3.2 showed how the species pool for Ontario in fact includes many quite different ecological regions. We observe that the vast majority of floral and faunal (and fungal) lists available to ecologists still use political boundaries. One still sees enormous numbers of maps of species distributions that stop at political boundaries. We urgently need to convert the political view of ecology to one based upon ecological regions. Many countries do have maps of ecological regions, which is a good starting point. Anyone who has tried to join maps of ecological regions from adjoining states or countries, however, will quickly find that often the maps do not fit together! That is to say, even maps of ecological regions are often confounded by political boundaries and different names.

Here is an example of the complexities. In southeastern Ontario, along the St. Lawrence River, there are boundaries among four different political entities: the nations of Canada and the USA, as well as the province of Ontario and the state of New York. In southeastern Ontario the southern deciduous forest region is mapped as part of Site Region 6E (Figure 3.2); Site Regions are now known as ecoregions. At the Canadian scale, this geographic area is designated as part of the Mixedwood Plains Ecozone and divided into three ecoregions: St. Lawrence Lowlands, Frontenac Axis, and Manitoulin-Lake Simcoe. One can find multiple maps for the adjoining state of New York, one of which federally designates the landscapes as Eastern Great Lakes and Hudson Lowlands (a Level III ecoregion) while another state-level map designates different ecoregions, one named the Great Lakes and the other named the St. Lawrence/ Champlain Valley. Fortunately, a binational Commission for Environmental Cooperation (CEC) has tried to bring some order to this situation, and has produced a continental atlas in which the entire area is one Level II ecoregion 5.2 "Mixed Wood Shield" (www.cec.org /north-american-environmental-atlas). Similar mapping and terminology problems likely await users in other less well explored parts of the world. Our point is not to make light of the complexities of ecological land classification, but rather to illustrate the challenges that face ecologists who wish to produce guides to species pools for particular ecological regions.

To put the example above into a larger context, the world map of ecoregions presented in Olson et al. (2001) is a relatively coarse-scale map, yet recognizes no less than 867 separate terrestrial ecoregions! And this is not counting marine ecoregions, which is a separate challenge: fortunately, there is now a similar map of marine natural areas, which is currently guiding the establishment of marine protected areas (Boonzaier and Pauly 2016). Since these global maps of ecoregions have a relatively coarse scale, there are smaller subunits, known as ecodistricts, nested within each ecoregion (e.g., the finer lines in Figure 3.2). We look forward to the day when there is a standard reference list for the species pool for each ecoregion, and eventually for each ecodistrict. This is an important goal, even if it seems difficult to attain. Otherwise, we will be stuck indefinitely with an unseemly mixture of manuals using political boundaries, combined with species lists from a haphazard selection of other geographic areas such as national parks. Order needs to be brought to this situation. This project would seem to overlap naturally with the continued identification of priority natural areas for a global protected areas system (Olson et al. 2001, Boonzaier and Pauly 2016). Superimposed on this exercise is the added challenge of identifying and exploring those areas that are still poorly known, like the Mekong River system mentioned earlier.

There is also a considerable need for consistent criteria for enumerating each pool. We have looked at some of the challenges above. Again, the project may seem overwhelming, but there are also precedents. A variety of codes for species nomenclature have brought some order to the older taxonomic literature, a literature that included duplicate names for species, and a literature in which species names were frequently assigned to minor variants and forms. Although the taxonomic literature is still evolving, and still being integrated with modern systematics, there is convergence on accepted species lists for the better-studied regions of Earth. For a popular account of the history of this endeavour, from the perspective of botany, but obviously with general lessons for all taxa, we refer you to *The Naming of Names: The Search for Order in the World of Plants* (Pavord 2005). We can anticipate that, through a similar process, standardized species pools can eventually be established for each ecoregion.

In the interim, we also note that species traits provide an alternative approach that may allow some progress in community ecology, independently of species pools. We explore this topic further in Chapters 6 and 8. Otherwise, we will close this chapter by repeating that in an emerging era of big data, we urgently need a global matrix showing species occurrences by ecological regions. This species pool by region data might look not unlike our Table 3.2, which offers such a provisional matrix for one geographic region back in 1958. This challenge allows us to acknowledge the contributions by preceding generations of explorers, collectors and systematists who have provided the foundations for this aspect of community ecology. Let us close with just one story – about Alfred Wallace (1823–1913), who explored the Amazon River in 1848 and published *A Narrative of Travels on the Amazon and Rio Negro* in 1853. Wallace and his companion, Henry Bates, collected specimens to document the species pool of this region. Their feats of physical endurance combined with daily struggles to protect their specimens during the expedition will resonate with anyone who has tried to collect original field data, and to protect it. Money was a problem and they were funded mostly by payment for specimens: Bates had an agent who gave him fourpence per specimen (from which he took 20 percent commission) so that Bates received a mere £27 for 20 months' work (Edmonds 1997). Yet, in spite of their exertions, Wallace was struck by catastrophe on the way back to England. "The ship on which he travelled caught fire

(August 6, 1852) and sank, with his entire magnificent collection and most of his journals, notes and sketches" (Mayr 1982, p. 418).

Our future knowledge of species pools rests on the shoulders of these early global explorers. Compiling authoritative lists of species pools for ecological regions will bring their adventures to a worthy conclusion.

KEY POINTS OF THE CHAPTER

- Species pools are the result of speciation and extinction, processes which mostly act at long time scales. We make the simplifying assumption to treat the pool as a relatively stable number.
- Standard reference works and classic natural history studies provide data on species pools at a variety of scales.
- Three types of pools should be distinguished: the *regional pool* (an unfiltered pool of species that takes into account geographic limits), the *habitat-specific pool* (a filtered pool that takes into account habitat limits) and the *potential pool* (which takes into account species that are adapted to different habitats and potential invasive species).
- A specified community contains a subset of the species in the pool, the actual number depending upon rates of input (immigration) and rates of loss (local extinction).
- There are levels of uncertainty in delineating pools, some of which arise out of life history traits, particularly those that affect rates of dispersal. Establishment rates can be enhanced through facilitation and reduced by competition.
- Many communities are not fully saturated because dispersal limitation limits the diversity of the community. In other words, there is often room for more species, but they just haven't arrived yet.
- Species pools can be quantified using two general approaches: species accumulation or downscaling.

4 Traits

Names and naming. Frank Rigler. Why species-based models are impractical for many purposes. Choosing traits. Some examples from birds and plants. Trait Matrices. Compilation and screening.

NAMING SPECIES

Most of us begin our study of ecology with reference to individual species. Someone has been kind enough to point out "That bird is a Blue Jay," "That animal is a White-Tailed Deer" or "That mushroom is a Chanterelle." We then begin to build up an understanding of the world based upon a species-by-species approach. This process begins very early in our life and is very helpful in kindergarten and public school. If you have ever had the good fortune to be in the field with an excellent bird watcher, or an excellent botanist, you will have discovered the pleasures of being able to name every single organism in your field of view. Anna Pavord (2005) has even written a delightful account of the search for order in the natural world, titled *The Naming of Names*. The first sentence is "Theophrastus is the first in the long list of men who fought to find the order they believed must exist in the dizzying variety of the natural world." Although this chapter in our book is titled *Traits*, we will begin our discussion by reviewing the costs and benefits of names and species in ecology. Some readers may be familiar with the argument, but others may benefit from a short review of some of the limitations of species-oriented biology. It is our experience that many people continue to do species-oriented ecology without considering its limitations.

It is useful, indeed necessary, to know the names of wild species, but it is possible to be out in the field with people who can name everything and yet otherwise know very little. In the bird community they are known as listers. Paul recalls people in the Nova Scotia Bird Society who had huge lists of birds, and travelled extensively to add bird names to their lists, and yet would not take the most minor steps to protect the habitats in which those wild birds lived. Back in 1976 the board of directors of this society actually refused to sign a letter objecting to the aerial spraying of forests with insecticides, even though this practice was devastating for wild birds in Nova Scotia. It was vivid proof that you can learn a name and yet know nothing.

When one leads guided walks through the forest, it is always important to do more than just give people names. Some of us have likely been on such excruciating walks where the guide thinks their only job is to teach us how to identify trees or mushrooms, without any further context. Ecologically, the names themselves mean little. We also need to teach what the organisms are doing, and why they occur where they do, and much more. For example, this is a Sugar Maple tree (its name). It forms a dense canopy, and all other tree regeneration then has to cope with shade and the understory flora consists of shade-tolerant species (its significance). Or similarly, this is a Balsam Poplar tree (its name). Balsam Poplar trees cannot regenerate in the shade, and are therefore dependent upon natural disturbances like forest fires or floods (its significance). It therefore has very small seeds that allow dispersal from one disturbed area to another (more on its significance). The point in these stories is not to get stuck at just teaching names, because they mean so little by themselves, and the names soon drift away from memory unless they are attached to additional knowledge to anchor them.

All the same, we do want to be clear: people do need to know the names of their local fauna and flora. What we are saying is that the name is not an end in itself, but just a first step to further knowledge. Nor are we using this chapter to make excuses for those lamentable ecologists who cannot name even the most common species in their own landscape. Naming organisms allows people in general, and undergraduates in particular, to first come to a primitive understanding of multispecies systems. Without a knowledge of names, we don't even know how many species there might be in a single lake or forest, let alone in the world. Just how many species are there, you ask? The *Catalogue of Life* website in 2018 (Roskov et al. 2018) states the number of described living species as 1,744,204. The relative abundance of the different groups is approximately as shown in Figure 4.1.

Our knowledge of Earth's biota is still incomplete. Hence, the precise size of the segments in Figure 4.1 is still unknown. There are particularly many unknown species in certain groups, like the insects, and new species are still being described daily. If you try to include these factors in the calculations, you end up with a much higher number of species on Earth (Mora et al. 2011). According to their work, on land, there are currently about 1.25 million species catalogued, with a total of 8.75 million expected. The difference is largely due to poor knowledge of animal species (934,434 described as opposed to 7,770,000 predicted), but also in other groups such as fungi (43,271 described as opposed to 611,000 predicted). Figure 4.1 already shows the vast preponderance of invertebrates on Earth, but the reality may be even more extreme than shown in that figure, in which three-quarters of all living species are invertebrates.

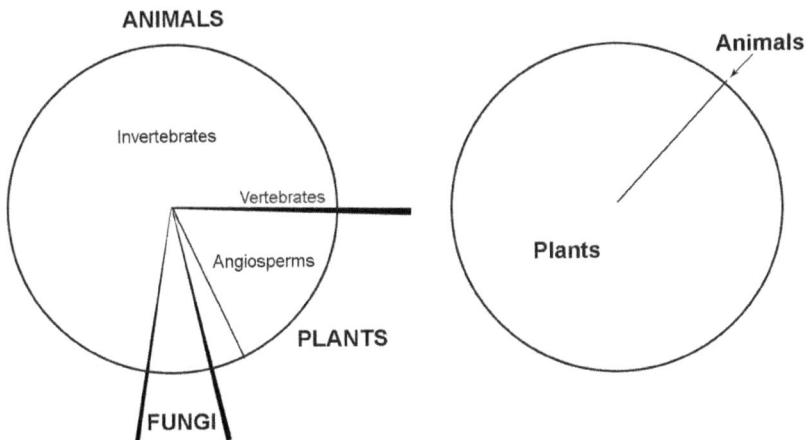

FIGURE 4.1 The relative abundance of species in the biosphere (left: with respect to numbers of taxa; right: with respect to their biomass), approximated. (After Colinvaux 1986.)

From the point of view of biomass, the image is dramatically different: from the biomass perspective, plants are the dominant life form. Indeed, they comprise the vast majority of the Earth's biomass, and much of this is one molecule, cellulose (Duchesne and Larson 1989). There are some uncertainties in interpreting the biomass data in Figure 4.1. Data on below-ground biomass is challenging to collect. This is true for both plants and animals. Across biomes, nearly half of the animal biomass occurs below ground (Fierer et al. 2009). Microbial communities also contribute to below-ground biomass: across biomes, microbial biomass is about 1 percent of the soil organic carbon. Microbial soil biomass is somewhere between 1 and 20 percent of total plant biomass (Fierer et al. 2009), a considerable margin of uncertainty.

Returning to the species perspective, there are further uncertainties. While it is entirely reasonable to ask roughly how many named species are on Earth, at a certain point, it is possible that new species of invertebrates may evolve faster than taxonomists can describe them. This is less of a problem with gymnosperms or large mammals. Unknown rates of extinction create further unknowns. However, for our purposes, as noted in the chapter on species pools, mostly we can treat the pool as fixed for the particular time and place in which we are working.

There is at least one more real risk to naming. Some biologists seem to get stuck early in their career, and focus on one species of organism for the rest of their professional lives. This may produce lots of papers, and lots of graduate students, but it does not necessarily advance our understanding of ecological systems in general, or ecological communities in particular. This illustrates how easy it is to slip into the species-by-species view of reality.

The species view still predominates in much of ecology. One critic of this approach to nature was Frank Rigler, a limnologist who, alas, is often overlooked. At the time, there was a driving need to understand lake ecology. There were multiple reasons for this. Pollution was becoming a problem in lakes, and it was unclear what was causing it. Lakes were being overfished or even fished out, and there were multiple questions about how to make fisheries sustainable. There were attempts to completely remake lakes by killing the fauna with rotenone, or other poisons, and then restocking with "desirable" species of fish. There was also the realization that lakes, being isolated systems, provided an opportunity to study a system with natural rather than artificial boundaries. In this sense, a lake is an island of sorts, unlike a tract of forest which may have no natural boundaries. Rigler saw vast amounts of money being spent on trying to understand the ecology of lakes on a species-by-species basis. Of course, there was lots of time and money being spent on individual species of fish, particularly fish that were popular with fishermen, like Lake Trout. But people constructing food webs for lakes were also studying individual species of plankton, and then trying to assemble these studies into whole-lake models.

Rigler's criticism of this approach is summarized in a 1982 paper titled "Recognition of the possible: an advantage of empiricism in ecology," published in the *Canadian Journal of Fisheries and Aquatic Sciences*, which is not likely prime reading for vertebrate or plant ecologists. Here is a key extract:

> A temperate lake may support 1000 species. If each species
> interacted with every other species we would have $(1000 \times 999)/2$ or
> 0.5×10^6 potential interactions to investigate. Each potential
> interaction must be demonstrated to be insignificant or quantified.
> If we estimate one man-year per potential interaction it would take
> half a million years to gather the data required for one systems
> analysis model. (Rigler 1982, p. 1328)

This statement clearly shows the limitations of species-based ecology. It provides a *reductio ad absurdum* for a great deal of scientific effort. The style of logic is to show that an assumption leads to a logically impossible or absurd result. In this case, Rigler argued that the species-oriented approach to studying lakes is intractable simply because of the time it would take to yield results. Time matters, particularly in a world where environmental problems have to be fixed in less than half a million years. We suggest that this single paragraph should be read several times by every practicing ecologist, who should then contemplate its consequences.

Part of the reason Rigler is overlooked is perhaps that many ecologists simply cannot conceive of an alternative research strategy. If you do not study species, how do you study ecology? Rigler did offer a solution. He suggested a greater emphasis upon empiricism, that is, the search for quantitative relationships between measurable dependent and independent variables. From this perspective, the study of ecology begins to look more like a large multiple regression problem. What are the key properties that we need to measure? How do we measure them effectively? What are the quantitative relationships between these properties? The classic example is the demonstrable relationship between plankton biomass and phosphorous concentrations in lake water (Peters 1992, Rigler and Peters 1995).

A further problem with the species-by-species approach is its lack of generality. However elegant and accurate, models based upon species have limited generality. As an illustration of this problem, try reading an article with multiple species names from a part of the world you do not know. If you live in North America, read a paper on the community ecology of trees in the Amazon, or birds in Australia. It will make little sense because all the names are unfamiliar. Even within a single system (wetlands), and with many years of experience, we have trouble reading papers on wetlands from different continents because none of the familiar reference points are there. Unless, of course, we add in some knowledge of traits. Oh, "that is a large tree with dissected leaves and buttress roots," or "that is a small bird that collects seeds off the ground." Once we add in some biological knowledge, we can process the names. Thus, models that are based upon species traits are more easily transferred from one continent to another. But that takes us back into needing to learn ever more natural history, which takes us right back to Rigler's challenge.

So, species can be a problem! One way to escape this problem is to focus on traits. We have become so accustomed to seeing the world through the lens of species that it is easy to forget that there is a complimentary, alternative view. Let us now look at the alternative.

A BIOLOGICAL PARADOX: TWO SYSTEMS FOR ONE REALITY

In order to understand biological reality, we must begin with a biological paradox – the terrestrial regions of Earth are indeed mostly covered in vegetation, but you need not one map, but two, to adequately describe the pattern. One map is based on evolutionary origins (phylogeny), while the other is based on evolutionary convergence (traits). Both maps took centuries to compose. We are going to focus on plants here, because they are foundational to all other groups of organisms, but we emphasize that the same logic can be applied to every other group of living organisms, from birds to rotifers. Explorers died while preparing these maps, including both Cook and Magellan. Humboldt and Wallace also had brushes with death but survived to old age. Still, you might think that with just one planet, one map surely would be enough to learn. It is not, and that is probably the most interesting background to the study of traits.

Most of you (all, we hope) will already be familiar with maps of the world's biomes. If you are not, it is time to dust off your favourite ecology textbook. Biomes are based mostly on convergence of traits, so in that sense they are on the useful track from the perspective of this chapter. But you should be aware of another much less familiar map. Figure 4.2 shows a map of the world's plants based upon phylogeny. Note how the evolutionary relatedness of plants reflects past events such as the breakup of Pangea and the formation of Gondwana. If you are trying to design a global reserve system to protect biological (i.e., evolutionary) diversity, you need such a map. Recent books describe in detail how to integrate phylogenetic relationships into ecology (Cadotte and Davies 2016, Swenson 2019). Evolutionary biologists and ecologists who study the relationships among traits across species acknowledge that species are not independent

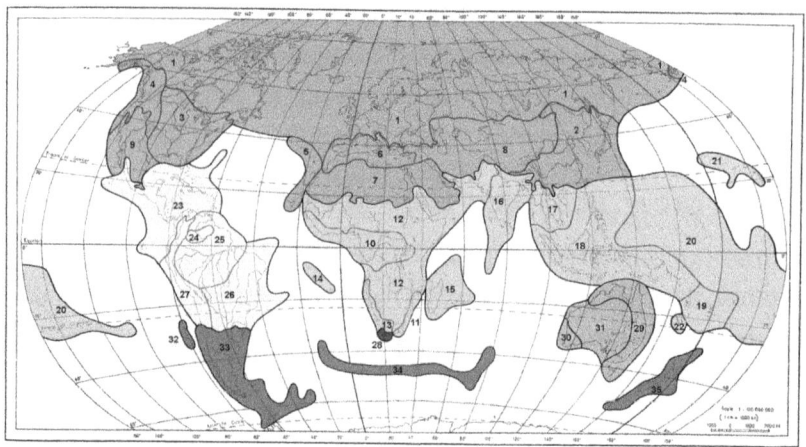

FIGURE 4.2 World floristic regions map the distribution of plants based upon phylogeny. Much of the world's plant diversity reflects long-term events such as continental drift and isolation of island fragments (from Keddy 2017). Such biogeographic patterns are very important for understanding evolution and for planning global systems for protected areas, but they are less useful for developing predictive models for community ecology.

replicates in a statistical sense. Species share an evolutionary history, and these relationships must be explicitly accounted for in our models. But when it comes to managing individual reserves or building models to guide management, you will likely need to know about the traits of the organisms you are working with.

Functional classifications are based on ecological traits. Rather than sorting individual organisms into groups that share a common evolutionary past, we could try to sort them into groups that share common ecological traits. The objective, then, is not to construct groups that reflect evolutionary divergence from a common ancestor, but rather to construct groups that reflect evolutionary convergence in response to common environmental pressures. It is based on the natural observation that some kinds of organisms look quite similar even though they have very different origins. Flightless birds are one example: kiwi (*Apteryx* spp.) do in some ways look like Greater Prairie Chickens (*Tympanuchus cupido*), albeit with very different bill shapes. Swimming animals are another: whales indeed look some-what like fish. The convergence of plants in arid lands is another example. While succulents occur in many deserts, deserts have cacti only in the Americas. This, for example, allowed "spaghetti western" films starring Clint Eastwood, allegedly taking place in the American West, to actually be filmed in Spain and Italy. Unless you are very observant, most plants of dry areas look superficially similar because they share certain functional traits. But as you learn more about plants, you start to notice films where the plants in the scenes contra-dict the alleged setting. Succulents, for example, all share certain traits for water conservation, but they are drawn from a wide array of plant families. If you know your desert plants well enough, you can tell when a western movie is really filmed in southern Europe or northern Africa. The same would apply if you saw a koala bear in an American western movie. Some producers know enough biology to avoid this problem, although it is amusing to see llamas appear in the 2004 film *Troy*, since visits by explorers to South America would not occur for several millennia.

FUNCTIONAL CLASSIFICATIONS: SOME HISTORICAL CONTEXT

As we look at the historical development of our understanding of traits, consider the situation of early global explorers who were faced with a bewildering array of plants in newly discovered parts of Earth. These explorers began with rather superficial but obvious traits (such as height and leaf size). Several hundred years later, we are slowly moving forward into more subtle traits (such as seed germination requirements or leaf physiology). One of the greatest obstacles to good functional understanding of plant and animal communities has been the lack of good data on traits. That situ-ation is now changing as large trait matrices are being con-structed. We shall have more about this later in this chapter. But we are also trying to tell a story about traits overall, so let us return to historical foundations.

One of the earliest functional classifications, prepared by Alexander von Humboldt, recognized 19 categories (Hauptformen). Most were named after a typical genus or family, such as Palmen-form, Cactus-form and Gras-form. As with many systems that are initially simple, Humboldt's was expanded to accommodate diversity and apparent exceptions. By 1872, Grisebach's work, *Die Vegetation der Erde*, included some 60 physiognomic types. These were grouped into seven main categories (Shimwell 1971, p. 67) that would still be familiar to most biologists today:

- woody plants (Holzgewächse)
- succulents (Succulente Gewächse)
- climbing plants (Schlinnggewächse)
- epiphytes (Epiphyten)
- herbs (Kräuter)
- grasses (Gräser)
- lower plants (Zellenpflanzen).

This system is really based upon a few simple traits such as plant height, presence or absence of woody tissue, leaf shape and leaf size. For early explorers, at the global scale, this was a start. The challenge has been to move forward to traits that are more subtle and offer more predictive power for community assembly. It has not been an easy task.

What are the traits that ecologists should focus upon? This is an important general question in ecology. In some groups of organisms, the important traits are obvious. This may be particularly the case with animals, where food is critical for survival and where mouth parts therefore carry a great deal of information about diet and habitat. This principle applies to both vertebrates and invertebrates, with invertebrates often being sorted by mouth parts and other factors affecting food collection and processing (Cummins 1973, Wong et al. 2018). Fungi, too, can be classified by the kinds of substrates they consume, and the kind of enzymes they use in the process (Allen et al. 2003, Treseder and Lennon 2015).

The task of compiling data on ecological traits is a problem in its own right. Exploring how these traits are related to environmental filters is also a problem in its own right. Whether we use traits to sort organisms into functional groups is a different but equally important problem, which we will discuss in Chapter 6. The point is that we need to keep these fields of enquiry separate in our minds, we need to see them as two or three independent problems, not one. Indeed, our historical example of introducing Alexander von Humboldt and August Grisebach actually illustrates the problem. We presented you with some functional *types* mostly without first addressing the *traits* that were measured. This illustrates the human tendency to muddle the two topics together.

We will start with an example from a well-studied group in the animal kingdom, birds. We will then turn our attention to plants, where the problem of finding traits is more difficult. We think the very difficulty of finding the right traits for plants actually presents an advantage: it forces us to think carefully about the kinds of traits we need to measure for any group of organisms, and the challenges of relating them to environmental filters.

MORPHOLOGICAL TRAITS IN BIRDS: BILLS ARE FUNDAMENTAL

There has long been a consensus that bill shape is an important trait in birds, because this trait (or more precisely, set of traits) determines the kinds of resources that can be consumed (Figure 4.3). Of course, this figure shows the traits being used to create functional groups, which again illustrates how the individual traits can be overlooked. Here we want to illustrate the link between bill shape and environmental factors, in this case, the relationship between bill traits and altitude. Altitude is a complex gradient that includes factors such as temperature, precipitation and habitat type, particularly the transition from forested to non-forested habitats.

To examine changes in bird traits with altitude, where better to begin than the Himalayas? The birds we will use are relatively small

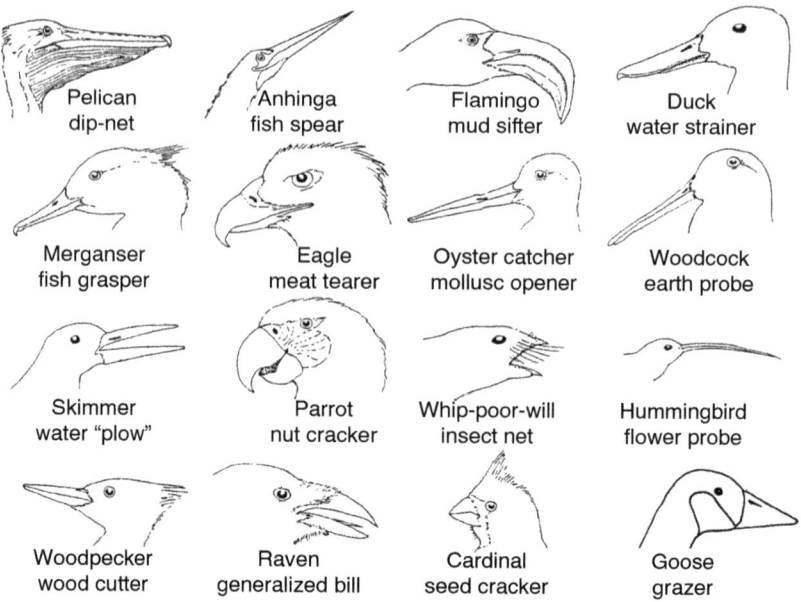

FIGURE 4.3 Bill shape is a fundamental trait for birds. (After Welty 1982.)

members of the Paridae, in which we find, in North America, the familiar Black-Capped Chickadee (*Poecile atricapillus*) as well as, in Europe, the familiar Great Tit *(Parus major)*. This family has long been of interest to ornithologists: David Lack (1964), for example, carried out a long-term study of population regulation in Great Tits in England, while MacArthur (1972) compared North American and European species of Paridae in *Geographical Ecology* (see figure 6.16, p. 158). Overall, there are about 55 species of Paridae found in the northern hemisphere, and the Himalayas are known to be an area of high species diversity in this group (Shao et al. 2016).

The set of traits measured on the Paridae of the Himalayas included bill length, width and depth, as well as more subtle shape measures (Shao et al. 2016). Other traits included body weight, wing length, tail length, tarsus length and culmen length. Three of these traits were significantly correlated with altitude (Figure 4.4). Figure 4.4a and b show bill characteristics: the difference between

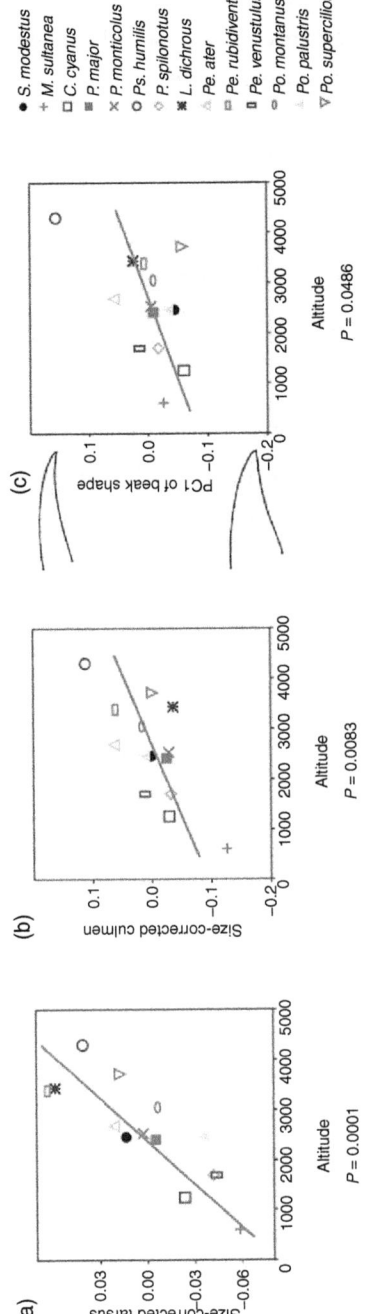

FIGURE 4.4 Traits of birds change with altitude in the Himalayas. The tarsus (a) is a leg bone, the culmen measures bill length (b) and the third trait (c) is a measure of bill shape. (After Shao et al. 2016.)

Legend:
● S. modestus
+ M. sultanea
□ C. cyanus
■ P. major
✕ P. monticolus
○ Ps. humilis
◇ P. spilonotus
✳ L. dichrous
△ Pe. ater
▯ Pe. rubidiventris
▯ Pe. venustulus
◔ Po. montanus
⊥ Po. palustris
▽ Po. superciliosus

(a)
Size-corrected tarsus
0.03
0.00
−0.03
−0.06
Altitude
0   1000 2000 3000 4000 5000
P = 0.0001

(b)
Size-corrected culmen
0.1
0.0
−0.1
−0.2
Altitude
0   1000 2000 3000 4000 5000
P = 0.0083

(c)
PC1 of beak shape
0.1
0.0
−0.1
−0.2
Altitude
0   1000 2000 3000 4000 5000
P = 0.0486

slender pointed and short blunt bill shape likely reflects the differences between probing on the ground as opposed to pecking in forested habitats. High altitudes, of course, have less forest. The increased length of the tarsus at higher altitudes is likely caused by the advantages of legs for hopping and walking on open ground.

This example, as noted, focuses upon a single group of birds, the Paridae. Larger data sets will make it possible to look for larger-scale patterns. Here are just two examples of how large data sets are emerging for the study of ecological traits: Renner and van Hoesel (2017) have provided a published data set on 24 ecological and functional traits for 99 common bird species in Germany, while Rodrigues et al. (2019) have produced a matrix of 44 morphological traits in 711 bird species found in the Atlantic Forests of South America.

We do not discuss some of the other aspects of this study, as we are focused upon one set of outcomes: correlations between traits and environment. Such data are frequently sandwiched into discussions of phylogeny, and more detailed analyses of changes in morphology within a single species. (Where is one less likely to find useful generalizations about traits and environment than when collecting data on single species?) In looking for simple, clear data on traits and environments, we have found that many of the available studies suffer from a shared set of limitations. First, the traits are measured within a single taxon, which minimizes any environmentally driven relationships between traits and habitat. Second, the empirical relationships are then accompanied by an array of speculations about the relative importance of phylogeny, convergent evolution and interspecific competition, and, of course, character convergence. The data offer no answers, nor can they. The quite remarkable thing about this process is the fact that it has no logical endpoint: one can measure and speculate *ad infinitum* without ever needing to come to a conclusion. It is entirely possible that we could wait another century and attend a future conference where someone has measured birds and speculated upon the relative importance of possible causes. Although Frank Rigler studied lakes, not birds, one can imagine him saying something acerbic that so long as there are enough birds, there will always be work for an ornithologist.

PLANT FUNCTIONAL TRAITS

Now we shall focus upon plant traits because this is where the challenge is arguably more difficult, and therefore has actually produced a good deal of careful thought (Westoby et al. 2002). There is also a thorough book devoted to describing plant functional diversity worldwide (Garnier et al. 2016). Unlike with animals, bill length or mouth size are not measurable options. Instead, all plants use a small set of resources (water, light and mineral nutrients), and it is therefore not possible to classify them easily on the basis of resources they use. Hence, ecologists have had to ask themselves carefully just which subset of traits should be measured. Weiher et al. (1999) suggest there are three main aspects of life history that need to be captured when measuring plant traits: dispersal, establishment, and persistence (Table 4.1). Note that they explicitly categorize some traits as "hard," meaning that although they are likely to be important, they are difficult to measure.

Let us consider mostly the first category, dispersal. All plants have to disperse to an appropriate site for regeneration. Indeed, it may be possible to classify plants by the type of regeneration site they use, hence the term "regeneration niche" (Grubb 1977). There are two distinct styles of dispersal, however: dispersal in space and dispersal in time. Dispersal in space has two obvious traits: the size of the seed and the presence of accessory structures that enhance dispersal. Consider seed size, or, more precisely, propagule size. Across the plant kingdom propagule size can vary over 10 orders of magnitude, from extremely small (ca. 0.0001 mg) to large (20 kg) (Moles et al. 2005). Some plant groups have minuscule seeds, known as dust seeds (Figure 4.5), making them quite dependent upon fungi for germination and establishment. Ferns and related groups of plants do not produce seeds, instead reproducing by means of spores. This trait, and the long-distance dispersal it makes possible, may be one reason for the persistence of non-seed-bearing plants in spite of the dominance of seed-

Table 4.1 *Three main aspects of life history that need to be captured in measuring plant traits, with hard and easy procedures for each (Weiher et al. 1999).*

Challenge	Hard trait	Easy trait
1. Dispersal		
Dispersal in space	Dispersal distance	Seed mass, dispersal mode
Dispersal in time	Propagule longevity	Seed mass, seed shape
2. Establishment		
Seedling growth	Seed mass	Seed mass
	Relative growth rate	Specific leaf area (SLA), relative growth rate
3. Persistence		
Seed production	Fecundity	Seed mass, above-ground biomass
Competitive ability	Competitive effect and response	Height, above-ground biomass
Plasticity	Reaction norm	SLA, LWC (leaf water content)
Holding space/ longevity	Life span	Life history, stem density
Acquiring space	Vegetative spread	Colonality
Response to disturbance;	Resprouting ability	Resprouting ability
stress and	phenology	onset of flowering
disturbance avoidance	palatability	SLA, LWC

bearing plants on Earth. At the other end of the size continuum are plants with large propagules, which often contain large seeds as well. Many tropical trees have large showy fruits, presumably associated with dispersal by large herbivores, but the upper limit is set by the Double Coconut (*Lodoicea maldivica*), a palm that is endemic to only two islands in the Seychelles, weighing in with 20 kg fruits (Moles et al. 2005).

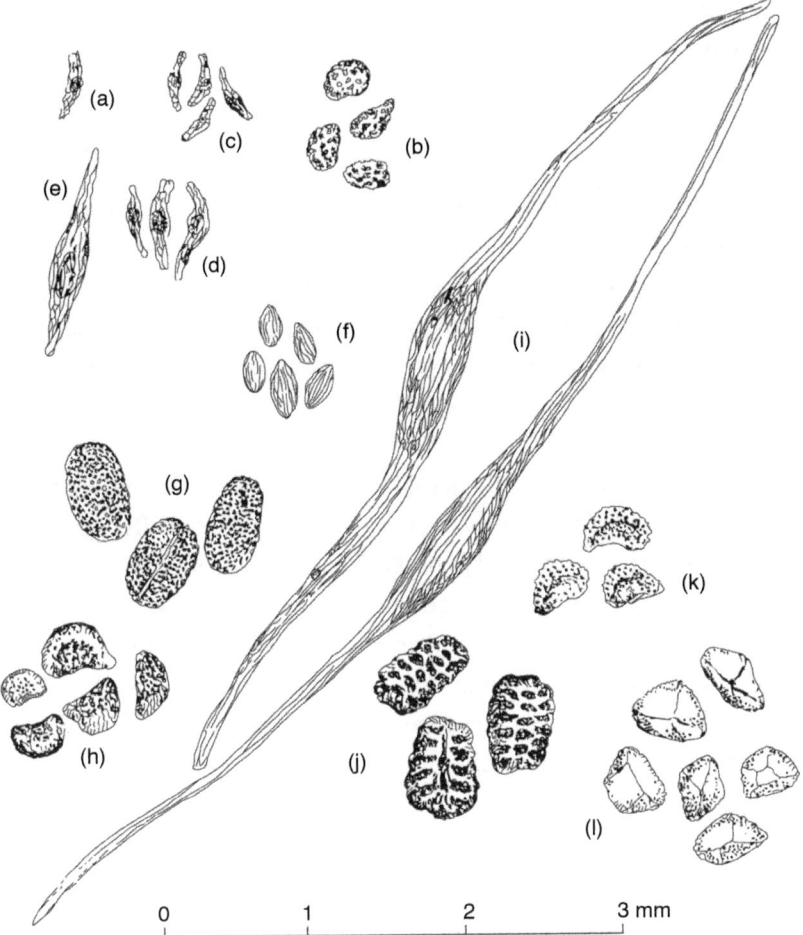

FIGURE 4.5 Very small seeds are known as dust seeds. While best known in the orchids, small seeds occur in a variety of plant groups. Here is a subset from the British flora: (a) *Gymnatoria conpsea* (an orchid); (b) *Orobanche elatior* (a parasite); (c) *Pyrola secunda*; (d) *Pyrola media*; (e) *Drosera anglica*; (f) *Crassula tillaea*; (g) *Digitalis purpurea*; (h) *Polycarpon tetraphyllum*; (i) *Narthecium ossifragum*; (j) *Scrophularia nodosa*; (k) *Spergularia rubra*; (l) *Samolus valerandi*. (From Salisbury 1942.)

The presence of special structures to aid dispersal is also significant. Dust seeds may be an exception of sorts, since their small size is an inherent adaption to wind dispersal. But even dust seeds may have other traits, such as long extensions, that likely enhance wind dispersal (Figure 4.5). Many such seeds are produced in capsules near the top of plants, enhancing the opportunities for wind to disperse the seeds. Then there is the wide array of other traits, such as various styles of parachutes that enhance wind dispersal and fruits that encourage animal dispersal. Standard reference works on botany will have illustrations and the accompanying nomenclature.

Dispersal in time is mostly associated with seeds that can persist during long periods of burial in the soil seed bank. The capacity to survive buried in the soil is often found in habitats subjected to repeated natural disturbances such as flooding, fire and wind throw. Survival as buried seeds is also important in deserts, where long years of drought are interrupted by occasional wet years (Pake and Venable 1996, Guo et al. 1998). Table 4.2 shows some data for the first group of species: those that withstand periods of flooding and regenerate during periods of low water. Persistence in soil is often associated with two morphological traits: small size and round shape (Thompson et al. 1993). Studies of seed germination in growth chambers have explored how long-term persistence and germination are related to environmental conditions such as the presence or absence of light, and the presence or absence of fluctuating temperatures (Grime et al. 1981, Shipley and Parent 1991). What other traits might be involved? Shipley and Parent (1991) measured three germination attributes (lag time, maximum germination rate, and final germination proportion) in 64 plant species. They found that germination responses were negatively related to relative growth rate (also measured in growth chambers) and positively related to the time needed to reach reproduction; surprisingly, seed size was not correlated with germination responses, at least in this sample of temperate zone wetland plants.

We suggested above that plant traits can be linked to three different aspects of life history: dispersal, establishment and persistence. In a collection of wetland plants, Shipley et al. (1989) found that juvenile traits, such as germination requirements, were uncorrelated with adult traits, suggesting that the dispersal and establishment

Table 4.2 *Many wetland plant species are capable of surviving long periods of time in flooded soil. The usual method of assaying for this trait is to count the number of seedlings that emerge from soil samples under suitable germination conditions (from Keddy 2010, where original references may be found).*

Wetland	Habitat	Seedlings (m^{-2})
Prairie marshes		
	Typha spp.	2,682
	Scirpus acutus	6,536
	S. maritimus	2,194
	Phragmites australis	2,398
	Distichlis spicata	850
	Open water	70
	Open water	3,549
	Scirpus validus	7,246
	Sparganium eurycarpum	2,175
	Typha glauca	5,447
	Scirpus fluviatilis	2,247
	Carex spp.	3,254
	Open water	2,900
	Typha glauca	3,016
	Wet meadow	826
	Scirpus fluviatilis	319
Fresh or brackish coastal marshes		
	Typha latifolia	14,768
	Former hayfield	7,232
	Myrica gale	4,496
	Streambank	11,295
	Mixed annuals	6,405
	Ambrosia spp.	9,810
	Typha spp.	13,670
	Zizania spp.	12,955
	Sagittaria lancifolia	2,564
	Spartina	32,826

Table 4.2 (cont.)

Wetland	Habitat	Seedlings (m^{-2})
Wet meadows in lakes or ponds		
	Waterline of lake	1,862
	30 cm below waterline	7,543
	60 cm below waterline	19,798
	90 cm below waterline	18,696
	120 cm below waterline	7,467
	150 cm below waterline	5,168
	Small lake, shoreline	8,511
	Small pond, sandy	22,500
	Small pond, organic	9,200
	Beaver pond	2,324

phases are quite distinct from those required for the adult phase of existence. Table 4.1 listed an array of traits that might affect the adult phase, under the category of "persistence." These traits include competitive ability, longevity and response to disturbance. Traits that defend plants against herbivores may be of particular importance (Pennings et al. 1998, Raven et al. 2005, Keddy 2017). We do not intend to review all such traits, but will mention here just two fundamental examples: leaf economics and water conduction.

THE FUNDAMENTAL GLOBAL TRAIT: LEAF ECONOMICS

If we were going to pick one trait to begin understanding the assembly of ecological communities on Earth, one could argue that it should deal with leaves. Leaves capture the sunlight upon which almost all living beings depend. We say almost all, because some organisms do not have leaves, but use photosynthetic stems, or microphylls, while other organisms living near hydrothermal vents do not depend primarily upon photosynthesis. These are, however, minority life forms. The fundamental importance of leaves was recognized even by early global explorers, but they had to focus mostly upon superficial aspects such as leaf size. Several centuries later, we have more powerful ways of thinking about leaves.

Let us start with leaf economics. Leaves have construction costs, not only the carbon, but the scarce nitrogen and phosphorus that are captured by roots. "Plants invest photosynthate and mineral nutrients in the construction of leaves, which in turn return a revenue stream of photosynthate over their life times" (Wright et al. 2004, p. 821). And there are trade-offs. Larger leaves may capture more sunlight but are also easily damaged. More stomata may allow more carbon uptake, but also may lead to more loss of water. New leaves are an investment in an uncertain future, and this investment is more costly in infertile soil. So how do these and similar trade-offs determine the kind of leaves one sees on wild plants?

On the topic of traits, where should we start in exploring leaf economics? Early explorers only had the time (and tools) to measure gross aspects of leaf morphology. Fortunately, we now have much more data on more subtle traits: an enormous study of leaf architecture used data from 2,548 species from 219 families at 175 sites (Wright et al. 2004). Let us consider some of the traits they measured:

1. The most basic property was LMA, or leaf mass per unit area. This is the dry-mass investment per unit of light-intercepting leaf area. Species with high LMA have a thicker leaf blade or denser tissue, or both. You can think of this as the investment in construction of the light-gathering apparatus.
2. Photosynthetic assimilation rate was also measured for each species, using high light, ample soil moisture and ambient CO_2 levels. Higher values of this trait correspond to higher stomatal conductance and/or higher drawdown of the CO_2 levels within the leaves.
3. Leaf nitrogen content was also measured. Nitrogen is essential for constructing proteins of the photosynthetic machinery, such as Rubisco, and a resource often scarce in plant communities.

Three other traits were also measured: (4) leaf phosphorus, (5) dark respiration rate and (6) lifespan. We shall focus upon just the first three here.

The database came from a global collaboration of physiological ecologists. In fact, this collaboration was one of the first of its kind in scale and scope, and signalled an emerging trend in ecology: the synthesis of trait data sets around the world to ask important questions at the global scale. If you measure such traits on large numbers of species, what kinds of patterns emerge? Figure 4.6 shows the relationships among these three traits. The assimilation rate is highest in plants with low LMA and high N content. The central axis can be thought of as a leaf economics spectrum.

This spectrum runs from species with potential for quick returns on investments of nutrients and dry-mass in leaves to species with a slower potential rate of return. At the quick-return end are species with high leaf nutrient concentrations, high rates of photosynthesis and respiration, short leaf lifetimes and low dry-mass investment per leaf area. At the slow-return end are species with long leaf lifetimes,

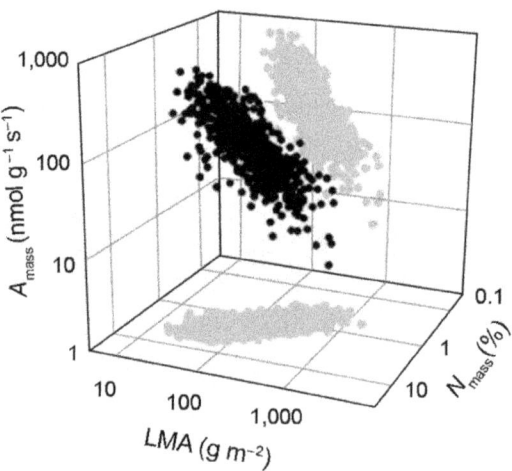

FIGURE 4.6 Three traits are important in leaf economics. Here are the relationships among leaf mass area (LMA), photosynthetic assimilation rate (A_{mass}) and leaf nitrogen levels (N_{mass}) in a set of 2,548 species of plants. (From Wright et al. 2004.)

expensive high-LMA leaf construction, low nutrient concentrations, and low rates of photosynthesis and respiration.

(p. 823)

To put this in evolutionary context, the earliest land plants like *Rhynia* did not even have leaves. The origin and diversification of leaves is likely best understood as the result of natural selection for increased ability to gather photons and carbon dioxide, but these are constrained by the costs of construction using scarce below-ground resources. This may be an opportune time to note that the acquisition of below-ground resources, in the vast majority of plant species, involves fungal symbionts. Mycorrhiza are the rule, not the exception (Brundrett 2009), and this suggests that types of fungi and types of functional relationships between roots and fungal hyphae are important traits to consider (Bergmann et al. 2020).

GETTING THE WATER TO THE LEAVES: TRAITS FOR WATER CONDUCTION

For leaves to photosynthesize, water is a key resource. This resource must be transported from the soil to the leaf by means of a continuous stream of water passing up the xylem. As tension within the xylem increases under water deficit, gas bubbles called embolisms form within the xylem and break the continuous stream of water. This blockage is called cavitation, and renders the xylem dysfunctional. The characteristics of xylem that determine susceptibility to cavitation vary greatly among plant species (Choat et al. 2012, Larter et al. 2015). Hence, this is an important trait to measure.

For many readers, some further explanation will be necessary. The general understanding of water transport in plants is as follows (King 1997): the evaporation of water within the leaves creates a water deficit in these tissues, ψ_x, which causes water to move into the leaves from the xylem. This water deficit pulls water from adjoining areas of xylem, and so moves water steadily upward in the xylem from the roots. In the roots, the water deficit within the plant moves water out of the soil and into the root. So, we may visualize the water within the xylem as a continuous thread, conducting the deficit in water potential in the leaves all the way back down to the soil.

A problem occurs when the water deficit is so great that trapped gas emboli cause cavitation in the xylem, breaking the thread, and ending the flow of water upward in the xylem. This is a particular problem for trees since the transport of water may involve movement upward for tens of meters. To some extent, the closing of stomata can help reduce low levels of ψ_x, but if soil water is not replenished, eventually xylem pressure will become so low that an embolism will form. The ability of species to survive and recover from drought is strongly related to their resistance to forming such embolisms, and this is largely determined by characteristics of the xylem (Choat et al. 2012, Larter et al. 2015).

A key trait is the vulnerability to stem xylem cavitation, defined as the xylem pressure leading to 50 percent loss of hydraulic conductance, or "P50" for short. For 480 species of trees, embolism resistance was greatest in trees growing in places having less than 1,000 mm mean annual precipitation (the world range is approximately 0 to nearly 5,000). In general, gymnosperms were more resistant than angiosperms (Choat et al. 2012, figure 3). Some of the lowest values in the data set – that is, species most resistant to embolism – were gymnosperms in the cypress family (Cupressaceae): *Actinostrobus acuminatus*, a desert shrub endemic to southwestern Australia (−14.10 MPa) and three species of juniper from southwestern North America, *Juniperus arizonica* (−13.30 MPa), *Juniperus ashei* (−13.10 MPa) and *Juniperus pinchotii* (−4.10 MPa).

To return to the implications of such data for the assembly of ecological communities, consider the distribution of coniferous trees in landscapes. Conifers (and more generally gymnosperms) originated in evolutionary history far before the flowing plants (angiosperms). The domination of Earth by flowering plants seems to have occurred after the impact of the asteroid that ended the Cretaceous and initiated the Tertiary, the same event that exterminated the dinosaurs.

Yet, although angiosperms are now dominant overall, gymnosperms are conspicuous in certain situations such as arid woodlands, boreal forests, and mountains. One conspicuous feature of gymnosperms is the lack of vessels in the xylem, a feature which may prevent freeze–thaw embolism, but Brodribb et al. (2012) explore some other traits that may also be important. Data in Choat et al. (2012) show that gymnosperms generally have a greater safety margin to cavitation than angiosperms. Thus, their occurrence in cold and dry locations overall may be attributed to embolism resistance. The most drought-resistant tree yet known is the gymnosperm *Callitris tuberculata* (P50 = –18.8 MPa!), which grows in dry areas of western Australia (Larter et al. 2015).

TRAITS AND TRAIT MATRICES

Good data on traits can be hard to find. This has been one of the major constraints for community ecology. This is not to say that our scientific literature is not full of observations on many species, it is just that the information has often been generated mostly haphazardly, species by species, or on small sets of species, and that it is consequently difficult to find reliable and meaningful data on large numbers of species. Often, the problem is not apparent until you actually take the time to go looking for it.

In Chapter 1, we introduced the concept of a trait matrix, Q species with K traits, and gave a primitive example in Table 1.2: one trait (survival time when flooded) measured for 23 species (trees). Now we introduce a more sophisticated example: 20 traits measured on 25 species of emergent wetland plants (Table 4.3). We introduce this example for at least four reasons. First, it is considerably larger than Table 1.2. Second, it contains a more complex mixture of traits: some are morphological (e.g., height of photosynthetic tissues), others are physiological (e.g., R_{max}, relative growth rate), others describe ecology (e.g., depth of rhizome burial), while still others are regeneration traits measured in growth chambers (e.g., percentage germination in light under fluctuating temperatures). A third feature of this

Table 4.3 *A matrix of 20 traits measured on 25 species of emergent wetland plants. The traits include morphological, physiological and ecological characteristics of the species. Trait columns 1–7 are for seedlings, that is, the regeneration stage. The eighth trait, growth rate (R_{max}) was measured on seedlings, but is generally considered a measure of growth rate in adults as well. Trait columns 9–20 are for adult plants. The names of the species and explanations of the traits are provided in the footnotes (adapted from Shipley et al. 1989, appendix 3).*

Species*											Trait**										
	sw	rlf	rlc	plf	plc	pdc	cn	R_{max}	vp	rt	sd	ld	rd	b	h	d	ct	gf	lt	lw
Ac	91	7.7	10.4	71	28	0	7.43	1.00	Yes	6.1	2.0	6.0	1.5	4.7	77.6	30.8	3	2	1.2	1.0
Apa	33	16.0	3.9	0	25	0	2.00	1.05	No	—	—	—	2.0	2.9	27.6	23.0	3	3	0.3	6.1
As	5	—	—	—	—	—	13.35	1.52	Yes	0.2	2.0	4.0	0.5	0.1	14.2	10.0	1	1	0.1	0.3
Cc	63	11.2	10.5	68	58	0	—	1.04	Yes	6.0	0.5	2.0	2.5	0.9	57.2	84.8	3	2	0.3	0.7
Cp	19	8.0	6.4	63	81	0	10.39	1.07	Yes	3.2	0.5	1.0	2.5	0.4	55.0	33.4	3	2	0.2	0.5
Cv	20	7.7	7.2	78	93	0	8.53	1.17	Yes	2.7	0.5	1.0	2.5	0.4	46.6	35.0	3	2	0.1	0.3
Da	53	6.6	9.8	71	82	0	6.94	0.89	Yes	3.0	3.0	5.0	5.0	0.8	48.0	20.0	2	2	0.1	0.5
Ee	31	12.9	10.3	16	4	0	4.79	—	Yes	0.7	0.1	4.0	0.5	0.1	36.0	0.1	5	2	—	—
Es	145	12.8	16.0	9	2	0	5.06	0.92	Yes	1.2	1.0	10.0	3.0	0.4	72.2	0.3	4	2	—	—
Ev	307	16.0	8.7	0	57	6	8.15	0.74	Yes	2.4	—	—	2.0	0.9	73.2	32.8	2	2	0.1	0.9
Iv	1,364	7.5	10.5	2	40	7	8.25	0.76	Yes	7.6	5.0	5.0	2.2	1.7	57.2	24.8	3	2	0.4	1.8

Table 4.3 (cont.)

Species*	Trait**																			
	sw	rlf	rlc	plf	plc	pdc	cn	R_{max}	vp	rt	sd	ld	rd	b	h	d	ct	gf	lt	lw
Jf	2	5.3	6.6	93	23	0	—	1.30	Yes	0.3	0.3	5.0	1.0	0.1	52.0	0.1	5	2	—	—
Ls	5	4.4	4.2	64	80	13	7.96	1.42	No	—	—	—	2.5	5.5	85.0	15.4	2	2	0.2	1.4
Lu	17	14.5	10.3	16	7	0	6.86	1.18	No	—	7.0	25.0	1.3	0.3	37.0	12.8	2	2	0.2	1.8
Pa	31	—	—	—	—	—	7.91	1.12	Yes	2.0	0.8	6.0	6.0	4.5	120.0	50.7	2	2	0.1	1.4
Pm	16	6.1	—	98	—	—	5.09	1.16	No	—	—	—	1.0	0.1	12.0	16.6	3	2	0.2	2.7
Ps	1	4.0	6.7	91	54	0	—	1.25	No	—	13.0	4.5	0.5	0.6	25.0	—	2	2	0.1	2.0
Rv	249	16.0	11.0	0	2	18	9.84	1.05	Yes	13.1	2.0	—	5.0	5.7	71.8	33.0	2	2	0.2	1.9
Sa	161	11.3	12.1	48	38	0	—	—	Yes	8.4	1.0	30.0	7.5	1.8	72.2	0.3	4	2	—	—
Sam	263	13.0	15.0	1	2	0	2.08	0.76	Yes	4.0	2.0	5.0	13.0	0.4	76.4	0.3	4	2	—	—
Sc	1	5.5	8.1	2	1	0	6.33	1.18	Yes	5.6	1.0	1.0	5.0	7.9	153.2	39.8	2	2	0.7	0.9
Se	5,204	7.2	6.1	30	0	0	5.83	1.16	Yes	4.7	21.0	27.0	—	13.7	100.4	47.0	3	2	3.6	1.0
Sl	29	9.8	10.0	20	10	3	6.74	1.21	Yes	7.1	25.0	25.0	5.0	3.2	51.4	37.6	3	3	0.4	3.5
Sp	161	7.2	6.1	30	90	47	16.62	0.77	Yes	2.9	1.0	11.0	3.0	7.8	161.8	42.0	2	2	—	1.2
Ta	4	3.3	5.6	66	85	0	6.89	1.63	Yes	14.0	4.0	25.0	10.0	20.5	167.2	49.2	3	3	1.8	0.7

Table 4.3 (cont.)

Code*	Species	Code**	Trait
Ac	*Acorus calamus* L.	sw	$\log_e$ of seed weight (g)***
Apa	*A lisma plantago-aquatica* L.	rlc	Average time to germination in the light under constant temperatures (20 °C) (days)
As	*Agrostis stolonifera* L.	rlf	Average time to germination in the light under fluctuating temperatures (10–30 °C) (days)
Cc	*Carex crinita* Lam.	plf	Percentage germination in the light under fluctuating temperatures (20 °C)
Cp	*Carex projecta* Mack.	plc	Percentage germination in the light under constant temperature (10–30 °C)
Cv	*Carex vulpinoidea* Michx.	pdc	Percentage germination in the dark under constant temperatures (20 °C)
Da	*Dulichium aurundinaceum* (L.) Britton	cn	Ratio of organic to inorganic fraction in seedling tissues (g/g)
Ee	*Eleocharis erythropoda* Stud.	R_{max}	Maximum relative growth rate of seedlings aged 10–30 days (g/g week)
Es	*Eleocharis smallii* Britton	vp	Presence of rhizomes or stolons
Ev	*Elymus virginicus* L.	rt	Thickness of rhizomes or stolons (where present) (mm)
Iv	*Iris versicolor* L.	sd	Shortest distance between successive shoots on a rhizome or stolon (where present) (cm)
Jf	*Juncus filiformis* L.	ld	Longest distance between successive shoots on a rhizome or stolon (where present) (cm)
Ls	*Lythrum salicaria* L.	b	Dry weight of above-ground biomass of a single ramet (g)
Lu	*Lycopus uniilorus* Mochx.	rd	Depth from soil surface to rhizome, stolon or roots (cm)

Table 4.3 (cont.)

Code*	Species	Code**	Trait
Pa	*Phalaris arundinacea* L.	d	Maximum diameter of the canopy of a single ramet (cm)
Pm	*Plantago major* L.	ct	Canopy type: leafy-monolayer (1), -multilayer (2), -graminoid (3); leafless–single stem (4), -tussock of stems (5)
Ps	*Penthorum sedoides* L.	h	Maximum height above ground of photosynthetic tissues of a single ramet (cm)
Rv	*Rumex verticillatus* L.	gf	Growth form: creeping (1), upright (2), rosette (3)
Sa	*Scirpus acutus* Muhl.		
Sam	*Scirpus americanus* Pers.	lw	Maximum leaf width (cm)
Sc	*Scirpus cyperinus* (L.) Kunth	lt	Maximum leaf thickness (mm)
Se	*Sparganium eurycarpum* Engelm.		
Sl	*Sagittaria latifolia* Willd.		
Sp	*Spartina pectinata* Link		
Ta	*Typha angustifolia* L.		

*** The following species had seeds too light to weigh accurately individually: *Agrostis stolonifera, Juncus filiformis, Lyrthrum salicaria, Penthorum sedoides, Scirpus cyperinus, Typha angustifolia.*

matrix is that it combines traits measured on both adults and juveniles, and appears to be the first test using Gower's similarity coefficients to show that one subsection of the matrix (adult traits) is statistically independent of another subsection of the matrix (juvenile traits). Finally, the authors used multivariate techniques to explore the patterns of variation within the trait matrix itself, an approach that allows possible generalizations to patterns that may occur in a much larger array of species. The authors conclude that in emergent wetland plants, juveniles range from species having small, rapidly germinating seeds with high growth rates to species having large, slowly germinating seeds with low growth rates. Adults range from large species apparently adapted to holding space, to smaller species that occupied gaps within the vegetation. The point is that there is a good deal of information we can extract by creating and investigating such trait matrices, even before we try to relate these traits to environmental filters.

Fortunately, trait matrices are now growing rapidly in both size and complexity. Trait matrices for local study areas are being published in journals at the rate of several per month. These individual matrices are being compiled into open-access global databases such as the TRY plant trait database (Kattge et al. 2020), and are being used to evaluate plant form and function at global scales (Díaz et al. 2016). Global compilation of disparate trait data sets is a difficult and expensive task, and the recent release of TRY version 5.0 was an extraordinary achievement. This is therefore an opportune time to note that we should distinguish between compilation and screening.

Compiled data are collected from pre-existing sources. The sources might be birds in museum collections, such as the matrix of bird traits published by Renner and van Hoesel (2017), which reports 24 ecological and functional traits for 99 bird species, or Rodrigues et al.'s (2018) compilation of 44 traits in 711 bird species. Another example is the book *Seeds* (Baskin and Baskin 2014), which compiles previously published studies of seed dormancy. Let us look at this latter project more carefully. There is a huge literature on the germination requirements for individual species, and these data can be compiled into one database and one book. Baskin and Baskin admit that they began studying seed germination on two species of plants in 1966, and, over the course of their careers studied and published many more. In 1998 they published *Seeds: Ecology, Biogeography, and Evolution of Dormancy and Germination*, which included data on more than 3,000 species. The book explores topics such as types of seed dormancy and effects of environmental conditions upon dormancy. The new edition (Baskin and Baskin 2014) has 1,600 pages and now offers information on dormancy on germination of more than 14,000 species. Such a book is a fine example of how compilations work. However, such compendia have their limitations, and it is those that we must think through here. Compilations may have huge amounts of information, but even if they are carefully indexed, problems can arise with their use in community ecology. "Why?" you may ask.

Imagine the task of compiling a trait matrix for a particular community. First, one needs to have a set of traits. This requires answering the sort of questions asked by Weiher and his co-authors: just what traits are needed for working with a particular habitat? For example, seed size is relatively easy to measure, but might not be all that useful for prediction. The response of seeds to fluctuating temperatures may be much more useful, but when one starts collecting information, it becomes clear that not all studies use the same range of fluctuating temperatures, or the same seed storage conditions prior to measuring response to temperature fluctuations. Or, some were measured in outdoor light, some in greenhouse light, and some in growth chamber light. Some were sown on soil in pots, others on filter paper in Petri dishes. And then, almost invariably, no matter how large the compendium, there are gaps. For working in a specific community we need trait data on all the species. In our experience, compendia are rarely available, and even when they are, some of the species that occur in the community are not included in the compendium. These data gaps rapidly become frustrating, and the number of gaps invariably increases as one tries to increase the number of species, and the number of traits, in the data matrix.

It is this sort of frustration that explains the importance of screening. Screening refers to the systematic measurement of one or more traits under controlled conditions in an entire set of species. Note that the definition of screening requires some sort of control of conditions, and, more importantly, often some deliberate manipulation of them. This is not data that can be collected from preserved museum specimens in bottles or trays, nor from dried plants stored in herbaria. This term is applied only when some manipulation is provided so that one is recording a comparative response of all the species to the same environmental factor.

One of the pioneers of comparative ecology, and hence of the technique of screening, was J. Philip Grime, working mostly at the University of Sheffield. In 1981 his research team systematically studied germination requirements for 403 species in the local flora in the Sheffield region of England. The seeds were all stored under similar conditions (dry, darkness, 5 °C). A chilling pre-treatment was also applied, using seeds stored in moist sand at 5 °C. Other work included testing for responses along a temperature gradient (from 5 to 40 °C). Seeds were also tested in light, dark and shaded conditions, as well as being exposed to temperature fluctuations of 15/20 °C. This work took six years of effort and provided information on an entire set of native species. Here it is useful to compare Grime's screening approach with that of Baskin and Baskin's compilation approach. Grime notes that the conditions may not have been optimal for all species, but all species received the same treatment and the conditions were therefore comparable. Note that Grime and his team could have published more than 100 individual papers tilted "The germination responses of species X to environmental factors A, B and C." To their credit, they did not. That would simply have bequeathed the task of compilation to a new generation of ecologists. Instead, they settled for one landmark paper in ecology (Grime et al. 1981).

In another case, an attempt was made to estimate the relative growth rate (RGR) for all species in this local flora (Grime and Hunt 1975). Now, RGR may seem like a complicated idea. The growth rate of a plant will change with environmental conditions, and different plants from different kinds of conditions. In order to measure RGR for a comparative data set, one standardized treatment has to be provided. In this case, RGR was measured as R_{max}, the maximum observed growth rate. As the authors note, "Ideally, to ensure measurement of the true R_{max} of which each species is capable, estimation should be made under the growth conditions optimal for that species. In order to find these optimal conditions, it would be necessary to subject each species to a wide range of factorially combined levels of the necessary growth factors" (p. 395). Of course, it is simply logistically impossible to provide those factorially combined factors, and one could argue that even if it were possible, each plant would be measured in a different set of conditions. Therefore, the environment used in screening for R_{max} had to be "a compromise between what was theoretically desirable and what was experimentally practicable." Certainly, it is important that the procedure be physically possible, but this statement may understate the importance that comparison under standardized conditions has its own value. After a series of initial experiments, R_{max} was measured in growth chambers with standard day length, artificial lighting, a temperature of 20 °C, a standard nutrient solution and so on, as described in that paper (pp. 397–400). This was the first such experiment on RGR for an entire local flora.

These two studies illustrate the general principles of screening. A large number of species is used. The conditions are standardized. This is a far different approach to so many of the papers one sees in ecological journals, where "the effect of light and mineral nutrients upon the growth of species X" provides data that are difficult to compile, even if, as is rarely the case, all use the same light, nutrients and growing conditions. Such studies illustrate in another way the problems that are created when one selects pairs of species from complex communities. Generality is difficult to find.

It is also possible to screen species by allowing them to interact with a standardized set of neighbours. This might allow one to assess traits such as resistance to herbivory, or resistance to disease, or resistance to neighbours. These kinds of comparative experiments that assess ecological traits by comparing performance to several selected species are generally known as "phytometer" experiments, to acknowledge that it was Clements (1933) who first used this term and suggested this kind of bioassay approach in ecology.

Let us consider the use of screening further, by exploring the challenge of measuring the relative competitive ability of plants. This trait was not measured in Table 4.3, although Shipley and his team suggested that some of the measured traits were measuring aspects of relative competitive ability. Hence, Gaudet and Keddy (1988) attempted to measure this trait more directly, by growing a set of 44 species individually in competition with two selected species. Hence, this was a kind of bioassay for relative competitive ability. Growing plants in competition with selected neighbours measures the effects of traits that might include size, RGR, photosynthetic responses to shading or canopy structure. In this specific example, two common wetland plants (*Lythrum salicaria* and *Penthorum sedoides*) were used in the bioassay of relative competitive ability. There are potential limitations: for example, it is possible that the results might depend upon which species of neighbour is chosen, soil conditions or even the growing conditions of a particular year. All such caveats are worth consideration. More importantly, they are not really caveats, but simply testable hypotheses. The important point is this: all 44 species experienced the same set of conditions and the same neighbours. Hence the responses are comparable. Indeed, this is why work using such data is often referred to as "comparative ecology."

Just how much do such methods change with environment or kinds of neighbours? Returning to those caveats (which are really just testable hypotheses), here are two examples of actual tests. Both deal with questions about how much consistency there is within competitive hierarchies. First, does changing the kinds of neighbours affect relative competitive ability within a set of species? Keddy et al. (1998) created seven different monocultures, each consisting of one of seven selected species of canopy-forming wetland plants. They then transplanted 48 different wetland species into such monocultures. When the rank order of competitive response in these 48 species was measured, it was similar across all seven monocultures. Hence, the competitive hierarchy is apparently rather consistent – at least across an array of 48 species growing under seven different canopies. Second, does changing the physical conditions affect relative competitive ability within a set of species? In another set of experiments, Keddy et al. (2000) examined the competitive hierarchy among 26 species of wetland plants under two contrasting fertility levels. The results showed that competitive performance was significantly correlated between the two fertility levels. Hence, it would appear from actual experiments using relatively large numbers of species (i.e., many more than two) that competitive hierarchies are relatively consistent. This is not to say, of course, that one could not find a pair of species that might reverse their interaction if one tested enough pairs of species. But we hope that we have already made the general point that testing general ideas using selected pairs of species is a remarkably inefficient (and quite possibly misleading) procedure for building knowledge about entire communities.

It is possible that such comparative methods can be used for many other traits and many other kinds of organisms, depending only upon the ingenuity of the experimentalist. So, there is ample opportunity for ecologists to be creative in investigation of traits that might be entered into trait matrices. The result of all these efforts of compilation and screening is a trait matrix, such as Table 4.3. The first dimension is the number of species, while the second is the number of traits. We mention this explicitly, since the challenge of building such matrices is to find the correct trade-off between the two. It is very difficult to do convincing studies in community ecology if one is missing some of the species, or if one is missing important traits. Standardized screening methods may allow us to build small matrices, and then extend them over time. But the challenge of ensuring that the data really are comparative argues for the importance of including the maximum number of species and traits from the start.

Where might we look for new kinds of traits? These will be based in the ecology and life history of each kind of organism. With birds, bills will always be important (Figure 4.1), probably followed by other traits such as body size, wing shape and leg length. With plants, regeneration traits will likely be important, as well as traits related to holding space and acquiring resources. Table 4.3 illustrates a principle that might be applied to many other groups of organisms: the importance of traits from a variety of life history stages. Trait matrices for amphibians could include columns for characteristics of the egg and larval stages. Trait matrices for corals might include columns for characteristics of the planktonic larvae. Trait matrices for dragonflies could include columns for characteristics of nymphs. Altogether, the commonest kind of trait matrices, compilations of traits measured upon adult specimens in museum collections, actually provide only a small subset of the functional traits that might be used in trait matrices constructed for community ecology.

How many traits do we need? Perhaps all organisms have a kind of "intrinsic dimensionality" (Laughlin 2014) arising out of the key filters that affected their evolution, in which case it will be possible to define inherent axes of trait variation. This will help guide the selection of traits and help ensure that all important axes of variation are included in matrices being assembled (Figure 4.7). Despite the very large number of traits that can be measured on species, the fact that traits exhibit correlations with one another suggests that trait variation among species can be adequately described using a reduced number of key traits (Laughlin 2014). The search for these key independent traits is currently underway.

If you are starting a new project in a local natural area and aim to use traits to understand the process of community assembly in your system, should you compile traits from the literature or screen traits yourself? It is tempting to try to skip the data-collection process by

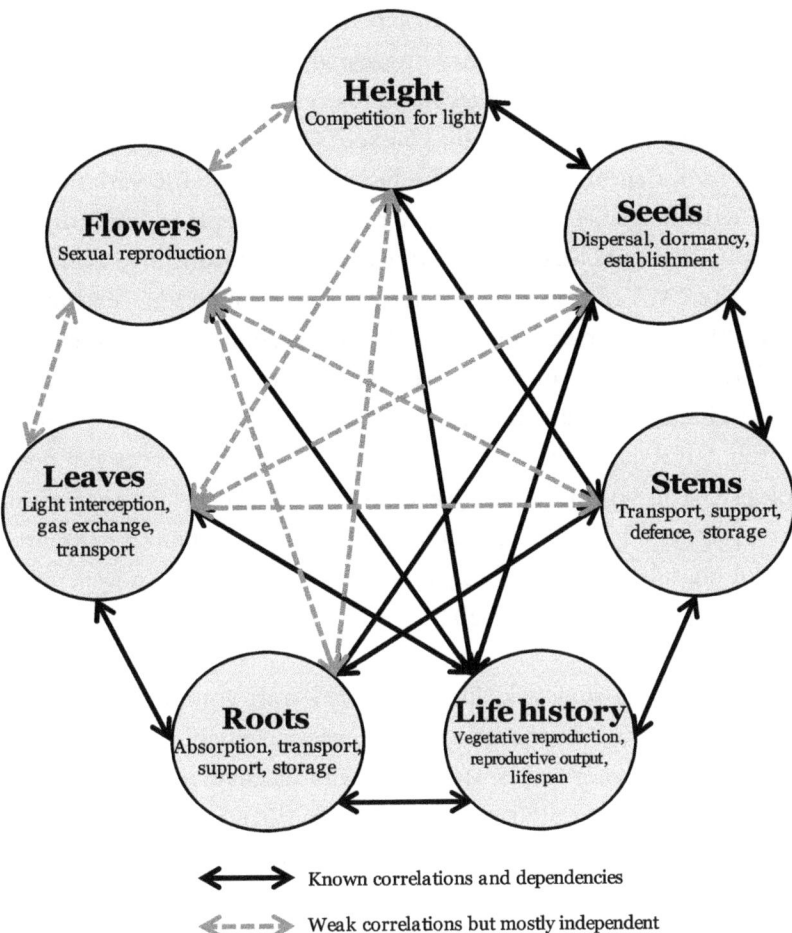

FIGURE 4.7 Although organism have many traits, these likely can be condensed into a smaller number of axes of trait variation. These define the "intrinsic dimensionality" of a group of organisms, and provide a kind of shopping list for the selection of traits. In this figure, at least seven different classes of traits are necessary (adapted from Laughlin 2014).

compiling traits for your species, especially if community composition data are already available, so you can jump straight to data analysis. However, we recommend that you screen the traits in your own system rather than simply grab traits of species from a global database. It may be too difficult to screen traits in standardized conditions the way that Grime has suggested, but measuring traits in similar local field conditions can be just as useful. There are many reasons to measure the traits for yourself. First, it will ensure that the traits of your species are accurate because intraspecific variation in traits within species can be large across the globe. It is doubtful, for example, that the LMA of *Poa pratensis* in your natural area is identical to the LMA of the same species measured on another continent under different climate and soil conditions. Second, it will ensure that you have no gaps in your data, because you measured traits on every species. It can be daunting to fill out a large species trait matrix from scratch, but it can be one of the most rewarding experiences as a field biologist. You will no doubt learn new things about each species in your system and it will force you to think critically about the importance of including each trait in your analysis. Third, your new screening of traits will then be of such high quality that it too can be added to a global database.

When, you may ask, are the global trait compilations to be used? Global trait databases are immensely valuable at the global scale. The leaf economics spectrum was discovered and popularized after global compilations of leaf traits (Wright et al. 2004). We now know that leaf economics traits are relatively independent of size-related traits, thanks to a global synthesis of the TRY database (Díaz et al. 2016). This synthesis showed us that the plants of Earth are arrayed on a largely two-dimensional constellation that appears much like our own Milky Way galaxy, determined by tissue quality and size. These studies have inspired ecologists around the world to synthesize other traits, including difficult-to-measure below-ground traits like roots. Through such a data synthesis, we have recently discovered that root diameter and specific root length (fine root length per unit dry mass) appear to represent a third dimension of functional variation among plants, which emphasizes the importance of collaboration with fungal partners for resource acquisition (Kramer-Walter et al. 2016, Bergmann et al. 2020). In short, global-scale compilations of traits are essential for answering global-scale questions.

TRAITS, ENVIRONMENTS AND APPLICATIONS

In this chapter we have directly pointed out one of the significant obstacles to developing predictive community ecology: the focus upon species names. We have drawn attention to the problem that it is therefore possible to become involved in research programs that have no logical conclusion, since the kind of model that is (wishfully?) being sought is based upon goals that are mathematically impossible to attain. That is the principal message of our small extract from Rigler's paper. Zeno's tortoise has returned in another guise. Moreover, research results that focus on species rather than traits cannot be easily moved from one geographic area to another. Figure 4.2 shows just how important it is that community ecology faces this problem squarely: there is simply no reason to expect a study carried out with species names in one biogeographic realm to be easily applied to another biogeographic realm. There are 35 such realms, and it hardly seems realistic to imagine a day with 35 individual branches of community ecology with its own scientific literature. Therefore, we have suggested the importance of focusing upon traits, and seeking out clear measurable relationships between traits and environments. To be clear, for managing a single park or reserve it may well be possible to approach the problem using species nomenclature. In certain cases it may even be necessary to focus on a small number of named species. However, this has an unfortunate consequence: it will then be difficult to apply the model to another protected area, except in the most general way. For example, can an alligator predator–prey model be extended to crocodiles, and if so, can it also be extended to cougars or wolves or trout? We do not expect you to answer that question, but we use it to make the point that we cannot envisage a future in which each protected area has its own set of models developed independently from other models in other reserves having other sets of species. Community ecology needs to transcend biogeographic realms, political borders and individual parks and reserves. In order to achieve this result, it is vital that younger scientists think carefully about Frank Rigler and Zeno's tortoise.

KEY POINTS OF THE CHAPTER

- It is important to be able to name the plants and animals in one's environment, but knowing the names does not in and of itself advance the study of ecology.
- Rigler argued that the species-oriented approach to studying ecology is intractable simply because of the time it would take to obtain enough information on each species to generalize to the community scale.
- Life on Earth can be classified in two complementary ways, using phylogeny and functional traits. Phylogenies describe the evolutionary relatedness among taxa, whereas functional traits describe important phenotypic properties of the taxa.
- Trait matrices provide the raw material for trait-based ecology. Functional traits reflect how filters have shaped the characteristics of living organisms over time and provide insight into how filters sort species into different habitats along environmental gradients.
- Compilations and screening are two distinct sources of data for trait matrices. Compilation of traits across studies is an important way of generating data for global-scale synthesis. Screening traits of local communities in the field or under standard conditions is the most effective way of generating quality data for local communities.

5   Trait–Environment Interactions

Three approaches to quantifying relationships between traits and environments. A fourth proposition. The fourth corner in community ecology. Community weighting of traits. A case study from a wetland in New Zealand. Population considerations. A message from Tansley.

THE POWER OF FILTERS

In Act I, Scene 3 of Shakespeare's *Macbeth*, Macbeth and Banquo cross the Caledonian heath on their way to Forres. On their journey they come upon three witches who predict that Macbeth will become King of Scotland. Banquo is quite skeptical of their predictive abilities, so he asks the witches, "If you can look into the seeds of time and say which grain will grow and which will not, speak then to me." Bill Shipley (2010) reminded us of this timeless verse, and so our intention in this chapter is to explore methods that allow us to measure how environments control the kinds of traits found in ecological communities so we can accurately predict "which grain will grow and which will not."

This is perhaps the most fundamental job for an ecologist. One of the greatest achievements of early naturalists was the detection of repeated patterns in ecological communities at a global scale. Consider deserts again. Despite virtually zero overlap in species composition across continents, well-travelled botanists discovered that the same functional types occurred in similar climates (recall Figure 2.2). For example, African Euphorbiaceae evolved similar columnar forms as the American Cactaceae, demonstrating that unrelated taxa have evolved similar traits to tolerate water limitation. Or consider another example, flooded areas, the opposite end of the moisture gradient. From the perspective of the Raunkiaer system, marshes are dominated by species that bury their meristems, whereas adjoining forests are dominated by species that keep their meristems held above ground. These examples of phenotypic convergence to solve a similar environmental problem across distant places illustrate the relevance of functional traits in community assembly.

The goal of this chapter is to review approaches that not only describe, but *quantify* the strength of trait–environment relationships. This leads us to our next proposition:

Proposition 4. Across multiple habitats, the power of a filter can be quantified by the strength of the relationship between traits and an environmental gradient.

Recall from Chapter 2 that we proposed that the power of a filter can be measured as the proportion of species that it removes from the species pool (Proposition 1). We now offer a second and independent method to measure the power of a filter, this time by using the patterns observed in multiple habitats arrayed along an environmental gradient. In this case, strength is measured by a standard statistical tool such as an R-square (e.g., you might look ahead to Figure 5.4). As we will see in Chapter 7, this is a necessary step toward *predicting* community assemblages. To put it another way, if we look at the variation in traits in an array of habitats, we can rank the factors in order of importance by measuring the proportion of the variance accounted for by each environmental factor. Our view is that these relationships are not only fundamental to community ecology, but essential for addressing practical problems in conservation, land management and restoration. To give an example of these practical problems, recall Chapter 2, where we explored how filters can cause habitat degradation in coastal wetlands, but how certain filters can equally be used as tools for wetland restoration (Figure 2.9). Overall, many conservation problems have a consistent theme: (1) the challenge of determining which environmental factors are currently having the most detrimental impact on a habitat; (2) the challenge of determining which environmental factors are the priority to manipulate for successful restoration (these may be identical); and (3) the challenge of finding the most efficient methods for applying these filters in the landscape. If we approach environmental problems without the intention to address the most important factors first, it is possible to waste enormous sums of money studying factors that are of only minor significance, rather like the horseman described by Stephen Leacock who "flung himself upon his horse and rode madly off in all directions." Our survey of filters in Chapter 2 therefore can be viewed as a kind of shopping list for priority factors that might be degrading natural areas, but that equally have the potential to be useful for restoration. Switching landscapes, now, from wetlands to arid lands, it would appear that grazing (particularly grazing by goats and cattle) is likely to be an important filter in many, perhaps most, arid landscapes. Returning to Proposition 4, note that we have now offered two methods for measuring the power of filters: one in a single habitat and one across multiple habitats.

OVERVIEW: QUANTIFYING TRAIT–ENVIRONMENT RELATIONSHIPS

We begin with the context: species pools **P** are filtered by the environment **E** based on how the traits **T** of species match the environmental conditions (Keddy 1992, McGill et al. 2006, Webb et al. 2010). That is to say, species with traits that are most adapted to an environment will perform better than a species with maladaptive traits, producing shifts in species composition across environmental gradients. The only way that traits can have an impact on the assembly of ecological communities is if traits are selected along the environmental gradient. As described in our fourth proposition, this selection of traits and filtering of the species pool is quantified by the statistical relationship between the trait and the environment. If the relationship between a trait and an environmental gradient is flat (i.e., the slope is indistinguishable from zero), then trait-based models will predict the same community composition in every habitat across the environmental gradient (Laughlin and Joshi 2015). In this case, the environment would exhibit no power at all as a filter. However, if the relationship between a trait and an environmental gradient is strong, then models predict different assemblages in each habitat along the gradient (Laughlin and Joshi 2015). Trait-based models of community assembly leverage these predictable trait–environment relationships to predict how community composition changes over space and time.

Dray and Legendre (2008) call the prospect of analyzing relationships between trait matrices and environmental data "an exciting challenge" (p. 3400). Pierre Legendre named this problem the "fourth corner problem," in reference to the lower right matrix in Figure 5.1 (Legendre et al. 1997). There are two general approaches for quantifying trait–environment relationships in this fourth corner: (1)

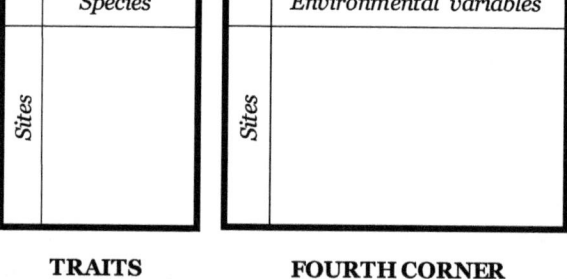

FIGURE 5.1 The three fundamental matrices of community ecology and the "fourth corner" matrix that is central to quantifying trait–environment relationships.

static abundance-based approaches and (2) dynamic fitness-based approaches (Table 5.1). A third approach might involve the use of common gardens, growing multiple replicate individuals of a species in multiple sites along an environmental gradient. As one example, in forestry, common garden tests of populations from different geographic origins started over 200 years ago. These "provenance" trials have been conducted for nearly all major commercial timber species in the northern hemisphere (Matheson and Raymond 1986, Matyas 1994).

STATIC ABUNDANCE-BASED APPROACHES

We quantify trait–environment relationships by using the community vector **C** as a snapshot of the community to look for patterns between the environment and the traits of species in the community. This is a snapshot because it represents an outcome: much, if not all, of the filtering has occurred in the past, so the observer is seeing the results of the filtering, not the process itself. We can observe the traits present in the snapshot, and by comparing this snapshot to others from other environments, we can explore the patterns among the species traits and the environments. The traits may, or may not, be weighted by the relative abundance of species in the vector.

The snapshot of the community vector of species abundances is the fundamental unit of community ecology, as described in Chapter 1. Whether it is a list of birds from a summer excursion, or a list of invertebrates within a scoop of marine sediment, ecology is full of lists of species abundances in particular locations. When trait data are available, other patterns can be explored. A growing number of studies have used species abundances and trait data to determine the strength and direction of trait–environment relationships (Thuiller et al. 2004, Ackerly and Cornwell 2007, Dray and Legendre 2008, Kraft et al. 2008, Thompson and McCarthy 2008, Cornwell and Ackerly 2009, 2010, He 2010, Pollock et al. 2012). Let us look at two of these in more detail. Of course, one could argue that many older studies going back to Raunkiaer also exemplify this approach, but we now have more extensive data and more refined statistical tools.

Table 5.1 *Comparison of the different approaches for quantifying trait–environment relationships (first two rows) and interactions (last three rows).*

Categories of approaches	Approaches	Response variable	Type of data	Biotic interactions	Data availability
Static abundance-based approaches	Community-weighted mean (CWM) regression	CWM traits	Static observational data	Cannot control for interactions	Abundant
	Fourth corner methods	Weighted CWM traits	Static observational data	Cannot control for interactions	Abundant
	Multi-level models	Species abundances or occurrence	Static observational data	Covariates to account for joint species occurrences	Abundant
Dynamic fitness-based approaches	Vital rate models	Individual growth, survival, reproduction	Longitudinal observational data	Covariate to control for neighbourhood competition	Infrequent
	Fitness models	Population growth rate	Longitudinal observational data	Low-density invasion growth rates	Rare

In a geographically extensive example, Thuiller et al. (2004) examined traits and environments in 88 *Leucadendron* taxa in the Cape Floristic Region. The genus *Leucadendron* is a group of evergreen shrubs in the Proteaceae endemic to South Africa, and this geographic region is a world biodiversity hotspot with a mostly Mediterranean climate (region 28 in Figure 4.2). There were more than 30,000 sites included in this study, although the traits included only morphological attributes. Species in dry Mediterranean climates had small leaves and small seed "cones," while species in the humid subtropical areas had large leaves with large "cones." (Since these are flowering plants, these latter structures are correctly called infructescences, true "cones" being restricted to gymnosperms).

At a much smaller geographical scale, Kraft et al. (2008) examined traits in 1,100 tree species in one 25-hectare plot of Amazonian forest. Although the area is much smaller than the preceding example, there were more species and a wider array of traits, including specific leaf area (SLA) and leaf nitrogen concentration, seed mass and wood density. In this case, topography was an important environmental factor: trees on ridgetops tended to have lower SLA, smaller leaves, heavier seeds and denser wood compared to adjoining valley areas. This result was an important early challenge to the null model that niches are unimportant in hyperdiverse tropical forests.

Static abundance-based approaches are the most common because these data are more readily available to ecologists. If you are interested in testing for such relationships, your goal is to acquire the three fundamental matrices of community ecology: a matrix of species abundances across a set of sites, a matrix of traits across all the species in the pool and a matrix of environmental variables across all the sites (Figure 5.1). With these three matrices in hand, you are equipped to explore the missing fourth corner.

MORE ON THE FOURTH CORNER PROBLEM

When you align the three fundamental matrices of community ecology, one notices that there is a "missing" matrix in the corner

(Figure 5.1). If this matrix could be filled, it would be filled by measurements of traits along environmental gradients, and this matrix could be used directly to quantify trait–environment relationships. The most common method for filling this fourth corner matrix is a "taxon-explicit" approach because it uses the matrix of species abundances to link the trait and environmental matrices together (Lavorel et al. 2008). This approach is static in nature because it is based on snapshots of species abundances, rather than population dynamics. These approaches can be sorted into three general methodological categories (Miller et al. 2018): CWM regression (Shipley et al. 2011), fourth corner methods and other weighted correlation metrics (Peres-Neto et al. 2017), and multi-level models (Pollock et al. 2012, Jamil et al. 2013).

Community-weighted mean (CWM) regressions evaluate the statistical relationship between traits and environments, where traits are computed as community-level averages weighted by the relative abundance of each species (Lavorel et al. 2008). CWM traits are computed as $\sum_{i=1}^{S} t_i p_i$, where S is the number of species in the community, t_i is the trait value of the ith species and p_i is the relative abundance of the ith species. CWM traits are then regressed on environmental variables in **E** to quantify the strength of the trait–environment relationship. This approach has been used widely to demonstrate which traits are most strongly explained by gradients in environmental conditions (Funk et al. 2017).

Let us consider an example from temperate rain forests in New Zealand (Simpson et al. 2016). CWM traits were calculated to evaluate how traits were related to both climate and soil gradients. The researchers assembled three matrices: a 324 plot by 64 species matrix of species relative abundances (i.e., the upper left matrix in Figure 5.1), a 64 species by 11 trait matrix of species-level average trait values (i.e., the lower left matrix in Figure 5.1), and a 324 plot by 4 environmental factor matrix of soil and climate variables (i.e., the upper right matrix in Figure 5.1). These matrices were used to compute CWM traits for each community, and these CWM traits were then modelled as functions of climate and soil. It was predicted that the effects of macroclimate on CWM traits would depend on the soil properties in the forest. In other words, Simpson et al. (2016) predicted that variation in traits could best be explained by interactions between climate and soil.

Specific leaf area emerged as a trait that was highest in phosphorus-rich soil within a wet and warm climate. In other words, forest communities in benign climates (wet and warm) and resource-rich soil (high phosphorus) were dominated by species with high SLA. This result makes sense because species with high SLA require more nutrients to fuel their faster growth rates in the wet and warm conditions. However, strong interactions between climate and soil complicated this general pattern, because forests in wet and warm climates in phosphorus-*poor* soil were dominated by species with low SLA! The relationship between climate and leaf traits depended on the soil properties, and this observation suggested that soil fertility imposed a stronger filter on species pools than climate. Species with high SLA were filtered out of sites with infertile soil even if the climate was benign. This example illustrates how CWM traits can be linked to environmental gradients and highlights the importance of considering multiple filters simultaneously.

The ease of computation and use of regression models has popularized the analysis of CWM traits, though it has come under recent criticism from both statistical and theoretical perspectives (Muscarella and Uriarte 2016, Peres-Neto et al. 2017, Laughlin et al. 2018b, Miller et al. 2018). First, from a theoretical perspective, CWM traits appear to be an unreliable metric for testing the adaptive value of a trait. We have assumed that CWM traits represent the optimal trait value in a community precisely because it is weighted toward the dominant species in the community (Shipley 2010). Logically, this argument makes sense: dominant species are evidently highly adapted to the local conditions. However, CWM traits cannot reliably be interpreted as optimal values for the given environment (Muscarella and Uriarte 2016), because dominant species do not always exhibit traits that conform to the mean. For example, dry tropical forests in Puerto Rico have dense wood on average, but some species that dominate dry sites have low wood density and cope with limited water availability by storing water in their stems and dropping their leaves in the driest season (Muscarella and Uriarte 2016). Multiple traits can override the importance of a single trait, suggesting that keeping track of the multidimensional phenotype is key to understanding the filtering of species along environmental gradients (Laughlin and Messier 2015).

Interpretations of CWM regressions did not consistently agree with a more direct analysis that examined the effect of traits on survival rates along an environmental gradient (Laughlin et al. 2018b). Consider again the relationship between SLA and soil fertility. Leaf economic theory predicts that in a soil with low fertility, perennial plant species with low SLA will have higher survival than species with high SLA because of their higher nutrient use efficiency (Chapin 1980). This is exactly what was observed in an analysis of survival rates among 46 perennial herbaceous plants in a Ponderosa Pine forest understory in Arizona (Laughlin et al. 2018b). We expected that these survival rates would be reflected in a positive relationship between CWM SLA and soil fertility – but there was no relationship detected. Why not? Possibly, dispersal limitation, recent disturbance, or rapid environmental change could prevent a snapshot measurement of CWM traits in a community from accurately reflecting the optimum trait value. If this is commonly the case, it might be a serious problem for community ecology as a whole. In what proportion of sites might traits be relatively unimportant in shaping the community vector, and what does this limitation mean for community assembly? These are open questions.

Second, from a statistical perspective, CWM regression suffers from poor type I error control, meaning that researchers will tend to find more significant relationships than actually exist. The culprit is the lack of independence of CWM values among plots that contain the same species (ter Braak et al. 2017, Miller et al. 2018). This has been confirmed by simulations that detected significant CWM trait-by-environment relationships even though no relationships were included in the simulated data (Miller et al. 2018). These results should make any ecologist pause.

There is a powerful and straightforward way to eliminate the problem of non-independence of CWM traits across sites because of shared species: fill the "fourth corner" matrix using a "taxon-free" approach, where traits are randomly sampled from a community without recording the species from which the sample was taken (Lavorel et al. 2008). Imagine a blindfolded student randomly sampling leaves from a forest. This random sample of leaves could be used to calculate an average leaf trait at the community level. The advantage of this method is that the trait measurements are completely independent from measurements at other sites. However, this method is rarely used in community ecology because there is a strong tradition of keeping track of the species from which the trait was measured given the interest in how traits vary across species. Also, calculating CWM traits is far easier by calculating an average trait value weighted by relative abundances! Given the popularity of species-based approaches, alternative statistical tests needed to be devised to account for this lack of dependence of CWM traits across sites. One such test is the "fourth corner" analysis.

Fourth corner analyses (Table 5.1) and other weighted correlation metrics are robust statistical approaches that address the lack of independence of CWM traits by assessing the significance of trait–environment relationships through permutation tests (Dray and Legendre 2008). These methods were proposed to solve the "fourth corner" problem, which refers to the nature of the matrix formulation to solve the methodological issues of linking species traits to environmental gradients (Legendre et al. 1997, Dray and Legendre 2008). The null model is that traits and environments do not influence community structure, so tests under this null hypothesis can be conducted by permuting species among sites or sites among species (Miller et al. 2018). Of course, this may seem like an extremely unrealistic scenario, but that is the purpose of a null model. Statistical tests of the null model repeat the tests of correlation after randomizing the location of species abundances in the matrix. The null model test asks: do we still detect a significant trait–environment relationship after randomly shuffling species occurrences among sites? If we do, then the relationship is not robust, it is due to the non-independence of shared species across sites. But if the null model tests of correlation are consistently worse than what is observed using the real data, then the relationship is robust. Fourth corner methods have considerable advantages because they reduce type I error rates and have more statistical power (Peres-Neto et al. 2017, ter Braak et al. 2017).

This method has been used in a variety of ecological contexts. For example, let us return to a theme from the preceding chapter, the traits of bird bills. In particular, are hummingbird bill traits significantly related to flowering traits of plants? Traits of bird bills may have evolved to specialize on specific flower shapes and sizes. Along an elevation gradient in neotropical rainforest near La Selva Biological Station in Costa Rica, bill length in hummingbirds was positively associated with corolla length in flowers, bill curvature was related to corolla curvature, and body mass was related to corolla volume (Maglianesi et al. 2014). These results suggest that bird species are filtered by floral resources through important trait–environment relationships (Figure 5.2). There is also an example of a fourth corner analysis from the Amazon basin, where traits and environmental data have been explored on 409 tree species from 53 forest plots (Fyllas et al. 2012). We will have a closer look at this study in the next chapter.

Multi-level models are the third approach for determining which traits are selected along environmental gradients (Brown et al. 2014, Laughlin et al. 2020). This approach differs considerably from the previous two approaches. Traits are no longer the response variable. The response variable is species abundances (or occurrences) and traits are moved to the other side of the equation and are used as predictors of species abundances. Rather than testing whether community-level trait averages are functions of an

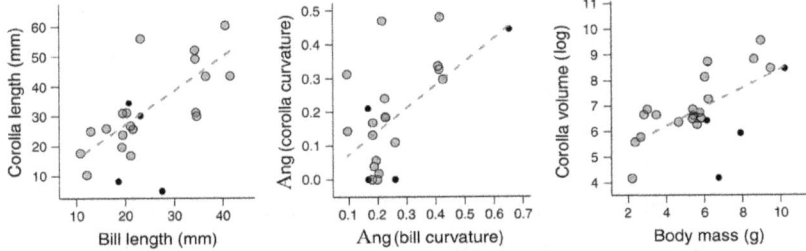

FIGURE 5.2 Three bill traits in hummingbirds are correlated with flower morphology. Each data point represents one hummingbird species at one elevation. (Simplified from Maglianesi et al. 2014.)

environmental filter, this approach asks whether the relationship between species abundances and environmental gradients depends on the traits of the species. In other words, it shifts the task from analyzing trait–environment *relationships* to trait–environment *interactions* that affect species abundances. It is, therefore, a more direct test of the adaptive value of traits because it tests whether an organism with a given trait is more likely to occur in one environment over another, all the while controlling for the fact that trait values are measured on species. What do we mean by this? The primary objective is to isolate the effect that traits have on the likelihood that an organism can survive in an environment. We saw previously that species can "get in the way" of analyses of CWM traits regressed on environmental conditions because the traits are measured on species that occur on multiple plots, thereby inflating type I error rates. How do we remove the "species effect" in order to isolate the pure effect that traits have in driving community assembly? Multi-level models control for the "species effect" by focusing on traits and environmental conditions as fixed effects while accounting for species as random effects.

Fixed effects are continuous or categorical predictor variables that are of primary interest to the researcher, whereas random effects are often used to account for the structure of the data. We refer the interested reader to an excellent text on multi-level models by Gelman and Hill (2006) for more information about the difference between fixed and random effects. In our case, traits and environments are fixed effects, whereas species are random effects. Typically, we already know that species abundances vary along environmental gradients, so we are not primarily interested in the main effect of the environment. These relationships have been known since the days of Humboldt. What we really want to know is whether the effect of the environment on the occurrence of an organism *depends on their traits*, so we include a trait–environment interaction as a fixed effect in the model (Laughlin et al. 2018b, 2020). However, the trait values are often measured at the species level. This means that for species i we use the same trait value t_i in every row of the data set where species i occurs. This does not have to be the case; some studies actually do have trait data on every individual (Worthy et al. 2020). But most don't. In order to control for the "species effect" we include species as random intercepts and allow their abundances to vary along the environmental gradient as a random slope (Laughlin et al. 2020). Put another way, the analysis makes two tests simultaneously, and therefore controls for as many confounding effects as possible: it quantifies whether traits affect abundances, given the environmental conditions, while simultaneously controlling for the undisputed fact that species vary along the environmental gradient. It controls for species to isolate the trait effect. This approach has been used to show that species with low SLA had positive responses to rocky soil (Pollock et al. 2012), and species with high SLA responded positively to manure application (Jamil et al. 2013).

AN EXAMPLE FROM A WETLAND

Let us now illustrate how each of these three approaches can be used to test the significance of trait–environment relationships. We will apply them all to one example: a kettlehole wetland on the South Island of New Zealand, the Wairepo kettlehole (Figure 5.3). Kettleholes are ephemeral wetlands, similar to vernal pools.

Kettleholes are important feeding grounds for the critically endangered Black Stilt (*Himantopus novaezelandiae*), the vulnerable Wrybill (*Anarhynchus frontalis*) and the declining White-Fronted Tern (*Sterna striata*). This round depression in the landscape was formed by glacial deposits 45 to 14 kyr ago and is nested in the middle of a valley of gently rolling hills surrounded by the Southern Alps. The flooding regime in the wetland is fed solely by rainwater and inundation occurs in winter when potential evapotranspiration is low and water levels often range from 20 to 60 cm deep. Traits were measured on individual plants that occurred along transects run from the highest elevation toward the lowest part of the wetland (Purcell et al. 2019). The data consist of percentage cover of 44 species in 187

FIGURE 5.3 The Wairepo kettlehole (a) is located in a protected Conservation Area managed by the New Zealand Department of Conservation. Plant associations that occur along the flooding gradient include environments: (b) frequently flooded sites dominated by *Eleocharis acuta*, (c) moderately flooded sites dominated by *Carex gaudichaudiana* and *Galium perpusillum*, (d) sites along the upper flood line dominated by grasses such as *Agrostis capillaris* and *Anthoxanthum odoratum* and (e) the upland dry sites dominated by *Pilosella officinarum* and *Leucopogon fraseri*. (After Purcell et al. 2019.)

quadrats, measures of flood duration in each quadrat, and two traits (root aerenchyma and vegetative height) for each species.

Should you wish to follow the process in detail, the code and data for all analyses in this and following chapters can be freely downloaded from https://github.com/danielLaughlin/CommunityEcology. The code depends on several R packages that you will need to have installed on your machine for the scripts to run properly. The three fundamental community matrices (Figure 5.1) that we will use in these examples are found in the Rdata file located in the Github repository. The percentage cover of 44 species across 187 quadrats are in a matrix called "comm. csv." Measurements of flooding duration (i.e., the average number of days the quadrat is flooded per year) at each of the 187 quadrats are in a matrix called "env.csv." Species-level averages of root aerenchyma and vegetative height for each of the 44 species are in a matrix called "traits.csv." We thank Bill Lee, an ecologist and evolutionary biologist at Manaaki Whenua – Landcare Research in New Zealand, for permission to make this data open access (Tanentzap et al. 2014, Tanentzap and Lee 2017, Purcell et al. 2019).

As we saw in Chapter 2, species that can produce aerenchyma in their roots are far more tolerant of flooded, oxygen-deprived soils because they can transport oxygen from their roots to their shoots to maintain respiration rates. We would therefore predict that root aerenchyma would increase toward the more frequently inundated locations in the wetland. Height is known as an important trait governing competition for light in wetlands (Keddy and Shipley 1989), and in plant communities overall (Grime 1979, Niklas et al. 1983). However, light availability does not vary along the gradient in this kettlehole, so in contrast to root aerenchyma we would not expect *a priori* that height drives the filtering of species along the flooding gradient.

For the first analysis, we compute the CWM values of aerenchyma and height for each of the 187 quadrats. Then we regress each CWM trait on flooding duration. As expected, CWM aerenchyma was positively associated with flooding duration (Figure 5.4a). In other words, communities located in frequently inundated sites were dominated by species with higher amounts of aerenchyma in their roots. Aerenchyma is an adaptive trait in waterlogged anoxic soil. However, CWM vegetative height was also significantly positively correlated with flooding duration, although the fit of this model was not as good (Figure 5.4b). Given the statistical problems associated with CWM trait–environment regression models, we need to check whether these significant correlations are robust.

For the second analysis, we can test the sensitivity of these CWM regression model results using the fourth corner method that computes the significance of the relationships through permutation against a null model. Recall that CWM regression suffers from high type I error rates due to the non-independence of CWM traits among sites that share the same species. Removing this species effect, the fourth corner analyses demonstrate that the relationship is still significant between root aerenchyma and flooding duration ($p < 0.022$; Figure 5.4a). However, the relationship between height and flooding duration disappeared under permutation ($p = 0.412$; Figure 5.4b). This suggests that root aerenchyma, but not height, is the more important trait that drives the filtering of species along the flooding gradient, and it illustrates the importance of permutation tests under the null model for testing trait-based hypotheses. This is also consistent with our understanding of wetland plants. The presence of aerenchyma might be expected to be directly related to flood tolerance. Height, on the other hand, might be expected to be a response to competition for light, which could be independent of the flood regime. That is, within any specified flood regime, tall plants will be an outcome in response to competition for light.

For the third analysis, we use the multi-level modelling approach to test the importance of each trait in community assembly. Rather than modelling average traits along the environmental gradient, we ask whether the occurrence of species along the flooding gradient depends on their traits (Figure 5.4c, d). To do so, we fit generalized linear mixed models using a binomial error structure and a log link function to model the presence and absence of species along

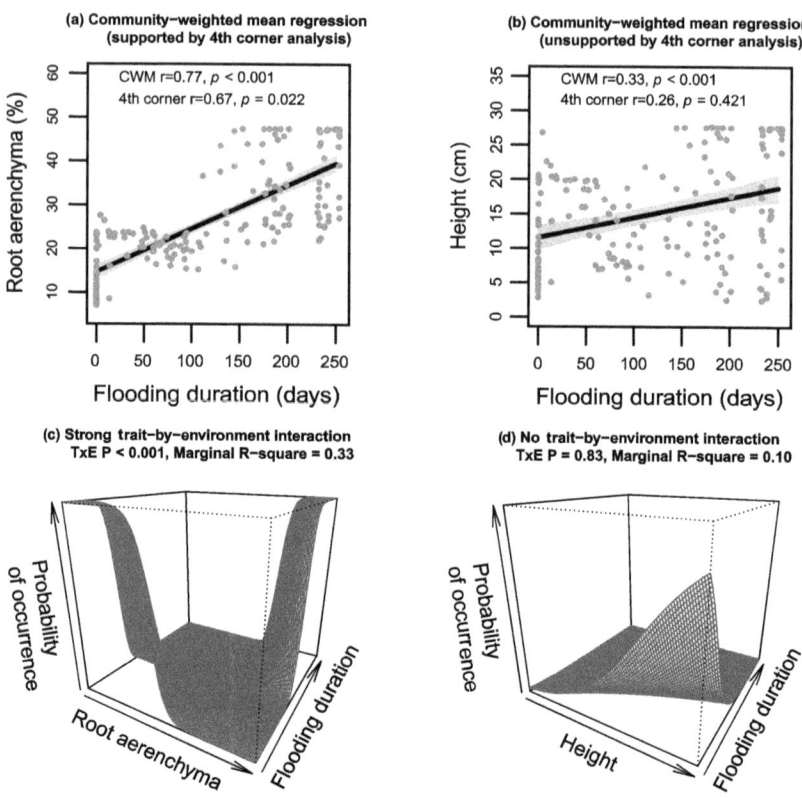

FIGURE 5.4 Abundance-based approaches were used to quantify trait–environment relationships in a wetland in New Zealand. The upper panels (a, b) illustrate CWM regression models and their statistical corrections using fourth corner analysis. The lower panels (c, d) illustrate the results of multi-level models that also confirm the relationships between root aerenchyma and flood duration. (Data from Purcell et al. 2019.)

the flooding gradient. In these models, the so-called fixed effects include the interaction between each trait and flooding duration. We use random effects to account for the structure of the data set. These include a random intercept for each quadrat to account for variation in occurrences across the quadrats. We also include random intercepts for each species, as well as random slopes with respect to flooding duration to account for the well-known fact that species occurrences vary along the flooding gradient.

The results of this model-based approach show a strong relationship between aerenchyma and flooding: species with less aerenchyma occurred in the dry end of the gradient and species with more aerenchyma occurred in the wet end of the gradient (Figure 5.4c). However, consistent with the preceding result, there was no significant relationship between height and flooding duration (Figure 5.4d). Thus, the results of the model-based approach agree with the fourth corner analyses.

DYNAMIC FITNESS-BASED APPROACHES

Up to this point in the chapter, we have been exploring methods for assessing the importance of traits in response to filters, using one particular kind of data: species abundances. The vector of species abundances **C** appeared in the very first chapter (recall Figure 1.1). Species abundances, however, are mere snapshots of the status of a species within a community. Communities are dynamic. Abundances of species change from year to year. It is now an appropriate time to ask: how useful are static snapshots of species abundances when assessing the role of traits in community assembly? Do alternative approaches exist that incorporate the dynamics of population size into our assessment of traits?

Population dynamics are determined by two fundamental components: stage-specific survival rates and stage-specific reproductive rates. These rates, when integrated, tell us whether a population is growing or shrinking (Merow et al. 2014, Ellner et al. 2016). It does not matter whether it is a population of bears, rotifers or corals. And, if a population is growing, on average, over several years within a community, we would consider that species to be particularly well matched to that environment. In contrast, if the population is shrinking then we would consider this species to not be well suited to that environment. Therefore, population growth rate would seem to be a useful measurement for determining how traits affect species performance in a given environment (Laughlin et al. 2020).

In life history theory and community ecology, population growth rate is a direct measure of fitness (Stearns 1992). The concept of fitness can be traced all the way back to Darwin's "struggle for existence," and fewer concepts in biology have received more attention. Or confusion. Here, we do not use fitness to refer to lifelong reproductive output because it is virtually impossible to measure on long-lived organisms (Harper 1977). Rather, we use the term fitness to denote the growth rate of a population. Species populations that are growing have positive fitness, and those that are declining have negative fitness.

Dynamic fitness-based approaches to assessing how traits affect performance in a given environment all begin with the assumption that species with negative population growth rates do not possess the right combination of traits (Laughlin and Messier 2015). This approach might seem inherently superior to a static abundance-based approach, so you might wonder why we did not start here. Well, population growth rates, and fitness more generally, are notoriously difficult to measure. Population ecologists spend years measuring detailed demographic responses of individual organisms in order to calculate population growth rates of species (Harper 1977). Usually such studies are focused upon single species. Data sets of population dynamics are only just beginning to be fully integrated with data on traits and environments to test theories in community ecology (Pistón et al. 2019, Kelly et al. 2021). The most likely reason is that in a world where systematic data on traits are challenging to find, systematic data on population response across an array of species are even more difficult to obtain. Fortunately, population ecologists are synthesizing population models and making them freely available to use for large-scale synthesis (Salguero-Gómez et al. 2015).

Although such data are rare, and may be difficult to obtain for a large number of species, in principle, the individual components of fitness, such as survival and growth rates, or reproductive rate, could be used to assess trait–environment relationships. For example, an analysis of survival rates in perennial plants demonstrated that the relationship between leaf traits and survival rates depended on soil fertility, documenting how specific traits mediate the filtering of species (Laughlin et al. 2018b). Individual rates of growth, survival and reproduction have also been related to environmental gradients in alpine plants in the Rocky Mountains (Blonder et al. 2018).

The analytical framework for detecting the effect of traits on population growth in a given environment is virtually identical to the previous multi-level model approach. Rather than using species occurrence as the response variable, it would be possible to use rates of survival or fecundity, or better yet, population growth rate. Unfortunately, we do not have data on individual vital rates or fitness for the wetland species in New Zealand, so we cannot apply this approach to our example data set used above. Overall, species abundance data are far more available than the repeated measurements data that are required to model the dynamics of population demography. It is also possible that species abundance data provide sufficient information for the purpose. It may be the case that the extra effort involved in collecting demographic data may not be worth it. Sometimes simple measures are entirely sufficient for advances in community ecology. Some discernment is necessary in choosing which variables are most appropriate, particularly in a world with competing demands for money and time.

This is perhaps a good time to emphasize the importance of discernment. Just because we can measure some attribute of a community, and just because we can model it with increasing sophistication, does not mean we should. Sir Arthur Tansley was not only a brilliant plant ecologist, but also an astute observer of human behaviour (he was a Jungian analyst as well as a plant ecologist). He noticed more than 100 years ago (Tansley 1914) that people have a tendency to measure things that are not necessarily that important: "The mere taking of an instrument in the field and recording of observations, or the collection and analysis of soil samples, is no guarantee of scientific results" (p. 200). Note that he wrote this at a time when there were no automated data loggers, no satellite images, no digital computers and no clouds for data storage. So, it is possible that the situation has become rather worse: just because you have a data logger and a helicopter and a team of graduate students and a huge grant, these do not guarantee that the measurements you are making matter one way or another. Again, Tansley's presidential address: "All science may be measurement ... but all measurement is not necessarily science."

CONCLUSION

In this chapter we have reviewed some of the territory in the fourth corner of community ecology: the relationship between traits and environments. Thus, we bring to an end one line of inquiry initiated in this book: the relationship between filters (Chapter 2) and traits (Chapter 4). In this chapter we have focused on the current state of quantitative methods for documenting trait–environment inter-actions. The scale of exploration has ranged across scales and habitats: from patterns in more than 80 species of shrubs in one genus arrayed across South Africa, to more than 100 species of trees squeezed into a single plot of Amazonian forest, to patterns in traits of 44 wetland plants in a New Zealand kettlehole. To appreciate the significance of these studies, it is, however, important to understand that this line of inquiry is rooted in questions that can be traced back through the history of ecology to pioneers including Humboldt and Raunkiaer.

At the same time, documenting the patterns in traits and environments is not an end in itself. Without discernment, it could turn into yet another decades-long excursion into multivariate descriptions, where studies of species in samples are replaced by studies of traits in samples. Will community ecologists approach this topic as a tool for prediction, or merely for more descriptions? In both cases the trait–environment relationships may remain the same, but the intent will to some extent determine the scientific value of the exercise. If the intent is simply to describe, then opportunities for making advances in predictive ecology are lost.

The study of trait–environment relationships is an intermediary exercise toward the goal of predicting community composition from species pools. This means that prediction of the community vector from the species pool may require us to have solid quantitative knowledge of the relationships between traits and environmental filters, but these relationships are best viewed as an intermediate step in a larger challenge.

In the following chapters, we will return to the problem of predicting the species vector. It is possible that this endeavour will be assisted by using traits to first sort species into functional groups. So, in the next chapter, we will give a cross-taxon overview of how functional groups are created from the information contained within trait matrices. It is entirely possible that community ecology actually has two fundamentally different lines of inquiry. Perhaps one line of inquiry explores how species are assigned to functional groups, while another explores how these functional groups are distributed among habitats.

Consider a landscape that illustrates these two aspects of community ecology – the south of Spain. This region is noteworthy for multiple reasons: it is a classic Mediterranean landscape (which we discussed in Chapter 3), it is a world biodiversity hotspot (the Baetic–Rifan complex) (Médail and Quézel 1999, Molina-Venegas et al. 2013) and it is an anthropogenic landscape that humans have manipulated and often degraded for over 30,000 years (Thirgood 1981). The lower mountains (Figure 5.5) have four common functional types of plants – evergreen trees, sclerophyllous shrubs, Chamaephytes and Geophytes – distributed along gradients including altitude, exposure and bedrock. The sclerophyllous shrubs provide a particularly well-documented example of evolutionary convergence (Mooney and Dunn 1970). Each functional group, however, contains so many species that it challenges both systematists and field naturalists. Does community ecology contain one set of rules for traits and functional types, and another set of rules for the redundancy with those types?

FIGURE 5.5 The garigue vegetation found at lower elevations of the Sierras de Tejeda, Almijara and Alhama Natural Park (southern Spain) has many kinds of shrubs with adaptations for water conservation. Traits include small, persistent and thickened leaves, aromaticity and thorns, as illustrated by (a) boxwood (*Buxus balearica*, Buxaceae), (b) rosemary (*Rosmarius officinalis*, Lamiaceae) and (c) spikethorn (*Maytenus sengalensis* subsp. *europaeus*, Celastraceae). (Photos by Cathy Keddy.)

To return to the opening of this chapter, and Shakespeare, it is possible that community ecology can be divided into two complementary subdisciplines: one dealing with the number of roles that will appear in the evolutionary play (trait-based functional types), and the other dealing with how many actors are qualified to play each role (species). It is possible that one set of actors are qualified to play the witches, whereas a different (and much smaller) set are qualified for the role of Macbeth himself.

KEY POINTS OF THE CHAPTER

- There are multiple quantitative tools available to quantify the relationships between environmental filters and traits. This general class of inquiry has been called the "fourth corner problem."
- Proposition 4: Across multiple habitats, the power of a filter can be quantified by the strength of the relationship between traits and an environmental gradient.
- One can measure traits at the community level with or without weighting them by species abundance.
- Most studies have used the vectors of species abundance C to assess the importance of traits along environmental gradients, but population dynamics can also be used.
- An ephemeral wetland in New Zealand is used to guide the reader through several ways of analyzing the relationship between traits and environments, and the data and R scripts for the analyses are available at https://github.com/danielLaughlin/CommunityEcology.
- A landscape often contains multiple functional groups, and within each functional group there can be many species.
- Documenting the patterns between traits and environments is not an end in itself. Without discernment, it could turn into yet another decades-long excursion into multivariate descriptions, where studies of species in samples are replaced by studies of traits in samples. Will community ecologists approach this topic as a tool for prediction, or merely for more descriptions?

6   Functional Groups

Assembly of furniture and ecological communities. Two stages of assembly. Phylogenetic as opposed to functional classification. Wetland plants. Tropical trees. Birds. Insects. Fish. Mammals. A pragmatic overview of clustering.

FURNITURE ASSEMBLY

Consider the task of assembling a bookcase from a box of parts: the task would be inordinately more complicated if we did not have a parts list. That is what the species pool provides for any community. But, following that same analogy, having the complete list of parts is still only one stage in the process of successful assembly. The parts need also to be enumerated and sorted before construction begins. For example, there will be just one back for a bookcase, but multiple shelves, and likely separate bags of screws and shelf pegs. So, all parts are not equal. How do we arrange the parts list (the species pool) for successful assembly (the community vector)? It is useful for the various parts to be separated into separate bags, one for screws, one for shelf pegs and so on. It is also often useful to know roughly what the piece of furniture will look like when assembled. You cannot build a barbeque out of the parts for a bookcase. This is why we need to explore functional groups. They provide a way of sorting the parts, and they may also help us define the target. So, let us start with our list of parts, the species pool and the trait matrix, and ask what we might learn from sorting the species into functional groups based upon their traits.

We should, of course, observe that working with functional groups might be just a tangent. It might be possible to find a tool that will allow us to leap directly from knowledge of the pool and the traits directly to the composition of the community. Just specify the filters and apply the appropriate mathematical procedure. Thus, we have procedures like maximum entropy and the Traitspace model (e.g., Shipley 2010, Shipley et al. 2011, Laughlin and Laughlin 2013), which we will explore in the next chapter. Before exploring these models, and their applications, it may be useful to consider an alternative approach to the problem of assembly. Perhaps it would be useful to break the process into stages.

Consider this possibility. We begin with the same data, the species pool and the trait matrix, but think about assembly in a slightly different way. We break it into two sub-problems. The first problem is to determine the functional groups that would be present in the final community. The second problem is to determine which species represent each functional group. Note that this nicely gives us a sense of natural hierarchy and scale: first the functional groups, then the species. In this chapter we will refer to this as the *nested approach*. That is, we will distinguish between the functional type outcomes from the species outcomes. Note that the species level itself could have two sub-goals, first the list of species and then their relative abundances. But, let us just focus on the first distinction: that between the functional group level and the species level. As mentioned above, it is possible, of course, that assembly models based on traits and pools can bypass this view entirely and take us directly from a trait matrix to a species outcome. Indeed, we shall see two such models in Chapter 7. Yet, even when this happens, it may be helpful to remain aware of the distinction between functional groups and species. We will focus mostly on functional groups in this chapter, and then revisit the topic in Chapter 8 (for a sneak preview, see Figure 8.1). There are several reasons for thinking about the assembly of functional groups as being somewhat separate from the assembly of species within functional groups.

- From the practical point of view, it is like assembling a piece of furniture. The many small parts actually belong to different sub-parts of the project:

how many drawers and doors are there in our bookcase? Will there be one, two or even three grills on our barbecue? Similarly, how many functional groups of plants are to be expected in a wetland? Or what functional groups of small mammals are likely to be found in a desert grassland?

- From a theoretical point of view, it has a certain kind of attraction, since the processes that determine the community seem to involve two rather different types. The first involves questions about how many niches or morphotypes there are in the presence of specified filters. The second question involves how many species occur within each functional group – how much redundancy there is in the system.

- This view also nicely accommodates a paradox. The paradox is that the accurate description of life on Earth seems inherently to require two contradictory systems: the phylogenetic and the functional. Phylogenetic approaches to biology focus upon delineating how many species there are and depicting their evolutionary relationships. Functional approaches focus more on the topic of convergence – that is, how many evolutionary solutions arise in particular habitats. Paul wrote a book chapter on this topic many years ago (Keddy 2010), based upon a talk in Japan which (he was later told) annoyed much of the audience because nearly all were trying to use phylogenetic data to assemble communities. Apparently, he should have been more diplomatic in pointing out the importance of functional data for community assembly, as some people are intellectually wedded to phylogenetic views of reality. As we noted in Chapter 4 in the section *A biological paradox: two systems for one reality*, it is a biological paradox of sorts that we need two entirely different maps, one phylogenetic and the other functional, to properly describe the distribution of Earth's biota. It may seem strange, but it is the nature of our biological reality, a reality which we cannot ignore or wish away. As F. Scott Fitzgerald wrote in an essay in *Esquire* back in 1936, "The test of a first-rate intelligence is the ability to hold two opposed ideas in mind at the same time and still retain the ability to function."

- There is a pre-existing set of statistical models for thinking about this problem in two stages: the old problems of combinatorics, the possible distributions of abundance of balls in a fixed number of buckets. One general conclusion arising from this field of statistics is that there is an enormous number of possible outcomes. If we can at least specify how many buckets exist – that is, how many functional groups species will be sorted into – it greatly reduces the number of possible outcomes.

- Then there is a convenient practical matter. For some problems, functional groups are a useful tool. If we know the pool, and if we have a trait matrix, then we also already have exactly the kind of data needed for sorting species into functional groups.

In this chapter, we are going to place our attention on the first problem – that is, the number of functional groups that may be found. If we assume that environments are acting as filters on the species pool, and selecting for certain traits, we may expect that filtering will lead to convergence in the traits of the species in a community (recall Figure 1.4). From this perspective, we might reasonably expect that all species occupying a community will be rather similar. Each habitat should have just one functional type that possesses the ideal traits for resisting the filtering effects of the environment. As to which species occur, we can assume that they are a subset of the list of species that belong to that functional type.

All too often, however, it seems that something is wrong with this thinking. Even rather small areas of habitat seem to support more than one functional type. Let us illustrate this with two contrasting habitats: those that are filtered by flooding, and those that are filtered by drought. Too much or too little water. Standing water is rather a severe filter, since few plants can tolerate the anoxic conditions of the rooting zone, as well as the mechanical stresses of water and waves. Yet, even in this extreme case, we find, to our surprise, several distinctive groups (Sculthorpe 1967, Hutchinson 1975). Some plants are erect and emerge above the water (think cattails), others have floating leaves (think water lilies) and others have entirely submerged leaves (most people don't even see these, but think seagrasses or certain pondweeds). As if this were not enough, there also may be plants that are entirely floating (duckweed). Hence, it would appear that even for a strong filter – flooding – natural selection has found four different ecomorphological solutions. We will return to this example below, but first let us consider the opposite extreme: deserts and drought.

The droughts typical of arid environments are also a strong filter. We can expect that species therein will have a common set of traits including reduced stomatal density and reduced surface to volume ratio (recall Table 2.1). Yet when we examine arid environments more closely, we find again that drought seems to have a number of different ecomorphological solutions. There is the obvious solution, epitomized by the cacti, with round or columnar plants storing water in the stem (Figure 2.2b). But, as Figure 2.2 showed, there are other options. There are thorny shrubs that produce leaves that are deciduous (Figure 2.2e), and there are plants that remain dormant as seeds until the occasional rainfall (pluviotherophytes) (Figure 2.2k). Apparently, there are no fewer than 12 trait solutions to the problems posed by drought! We doubt that any of us would have imagined this outcome if we had made an *a priori* list of possible types of plants in response to drought.

These two examples pose a fundamental challenge to our thinking about filters and traits altogether. It seems that there is no automatic correspondence of one filter to one trait. For some reason, nature does appear to offer multiple solutions in some, perhaps all, cases.

This leaves us with two possible challenges. We may still try to come up with a method for predicting the number of functional types that can arise in an environment, *a priori*, from knowledge of traits and filters themselves. Yet, given the two examples above, flooding and aridity, it would seem that predicting the number of functional types from first principles will be a considerable challenge, and possibly well beyond anything we can expect to achieve in the short term. Fortunately, we have a second option, a more pragmatic approach. That is the approach we will focus upon here, while not discounting that it may eventually be possible to achieve the more difficult goal of *a priori* prediction.

Suppose we decide to accept that evolutionary solutions are surprisingly unpredictable. We may not be able to predict from theory how many life forms will arise in a particular location. But we can measure them. We may need to accept that the number of functional groups in a habitat is simply something that, at least in the short run, one has to measure. To take an analogy from physics, instead of trying to predict from first principles how many elements might be found in the cosmos, we instead accept that we need to compile a periodic table of some sort from a large number of observations. For each specified habitat type and taxon, then, we would need to specify how many functional groups can occur. Many scientists have already done so for a particular group of species or a particular location.

Enough general principles. Let us turn to some examples to see how several generations of biologists and ecologists have dealt with this problem.

FUNCTIONAL CLASSIFICATION FOR ECOLOGICAL PREDICTION

As we have already explained, classification into groups could have two objectives: (1) forming groups with similar evolutionary histories in order to reconstruct phylogenies, or (2) forming groups with similar ecological traits for predictive ecology. The former approach has had a major impact upon the historical development of ecology: many of the most high-profile research questions in ecology dealing with diversity (e.g., Hutchinson 1959, Connell 1978, Huston 1979, May 1986) focus upon species, that is, upon a phylogenetic approach. The recent proliferation of molecular approaches to systematics has greatly reinforced this view of nature, sometimes to the detriment of functional thinking. If we begin with phylogenetic species classifications, we naturally fall into a certain line of inquiry, with its own view of communities. The logic appears to go in the following manner. Since there is a large number of species, how did they all arise? Darwin provided an answer, and stimulated a century of research into the mechanisms and consequences of evolution through natural selection. This leads to the second major question: how do all these species coexist? The coexistence of many different species is the great question bequeathed by Darwin's work on origins. Coexistence has therefore been a central theme of ecology, with two landmarks being Hutchinson's 1959 paper entitled "Homage to Santa Rosalia" (Jackson 1981, May 1986), and the "Paradox of the plankton" (Hutchinson 1961). This naturally leads to the question of similar species, and how they manage to coexist. A familiar historical example is MacArthur's investigation of five coexisting warblers (1958): "The species are congeneric, have roughly similar sizes and shapes, and all are mainly in insectivorous" (p. 599). Only a few years later, there was a long series of numbered papers sharing the title "The comparative biology of closely related species living in the same area" by John Harper and his students (e.g., Clatworthy and Harper 1962, Harper and McNaughton 1962).

If, however, we begin with functional classifications, the path of inquiry has a different logic. While there is of course enormous species diversity of the biosphere, there is also obvious repetition of certain themes. Most wetlands around the world will have groups of floating-leaved aquatics, wading birds and predatory insects, but the names of the species change with geography. From the functional point of view, there are at least three important questions: (1) What are these major convergent groups? (2) How many groups are there? And, perhaps, (3) how many groups do we need for a sufficient level of precision in our ecological models? Growing out of this are other questions. What are the traits that they share? How do we use a knowledge of these traits to predict how a particular functional group will change after an external perturbation? How can we use a knowledge of these traits to predict the group of species that will be present in a specified environment?

So, we intend to explore the process of creating functional classifications from traits. Often, the search for functional groups has been approached somewhat haphazardly, with separate lines of inquiry carried out quite independently for specified groups of organisms within haphazardly selected habitats. Thus, the same concepts may have different names, depending upon the organism being used, which seems quite unnecessary. For example, plants have *life forms*, insects have *feeding groups* and birds have *guilds*. Here, we are trying to look beyond the use of language to the search for foundational types of organisms based upon similar sets of traits. Regardless of the language in the particular papers, we will use the general term *functional type*. We also note in passing that this redundancy in terms can create its own sub-industry of publications, such as, is a life form (plants) the same as a guild (birds)? These sorts of questions can be asked for each pairwise combination of terms. We want to avoid fuelling this sort of debate, and instead stay focused upon the central question: whatever the group of organisms, how do we sort them into groups based upon their traits? Now some examples.

FUNCTIONAL GROUPS IN PLANTS

We have already mentioned how Alexander von Humboldt's expeditions back in the 1800s led him to attempt to sort plants into types: his groups included woody plants (Holzgewächse), succulents (Succulente Gewächse), climbing plants (Schlinngewächse), epiphytes (Epiphyten), herbs (Kräuter), grasses (Gräser) and lower plants (Zellenpflanzen).

Our next example is the Raunkiaer system (Figure 6.1) which remains relevant because, as we outlined in Chapter 1, it is simple, it uses one ecologically defensible trait and it is still widely used. The principle trait is the location of perennating meristems. Meristems (buds) are the basic units by which plants grow. They may be temporary or long-lived, and above or below ground. Indeed, you can argue that most plants are colonies of many meristems, in which case we can draw an analogy with colonial animals like corals. Focusing upon how plants protect meristems seems an eminently sensible way to categorize them, particularly when you are comparing plants from widely different environments. To apply the Raunkiaer system, we must focus on only a subset of these meristems – those that overwinter, or, to use a more general term, perennate. It is common to think of perennating meristems as the overwintering meristems, but in warmer parts of the world they may equally tolerate periods of drought, or allow regrowth after fire or grazing (Ott et al. 2019). Raunkiaer appreciated that the location of perennating meristems was closely connected to climate, easily recognizable in the field, and a single characteristic that could be enumerated for statistical comparisons of geographical locations (Raunkiaer 1908). There are five main categories.

1. *Phanerophytes* produce their perennating meristems high above ground. You know them as trees or shrubs, and they are put into four quite artificial categories according to height.
2. *Chamaephytes* are also woody, but their perennating meristems are produced close to the ground, and in colder climates are often buried by winter snow.
3. *Hemicryptophytes* produce their meristems at the soil surface.
4. *Cryptophytes* produce their meristems beneath the soil surface or underwater.
5. *Therophytes* are annuals, and pass unfavourable periods as embryos within a seed.

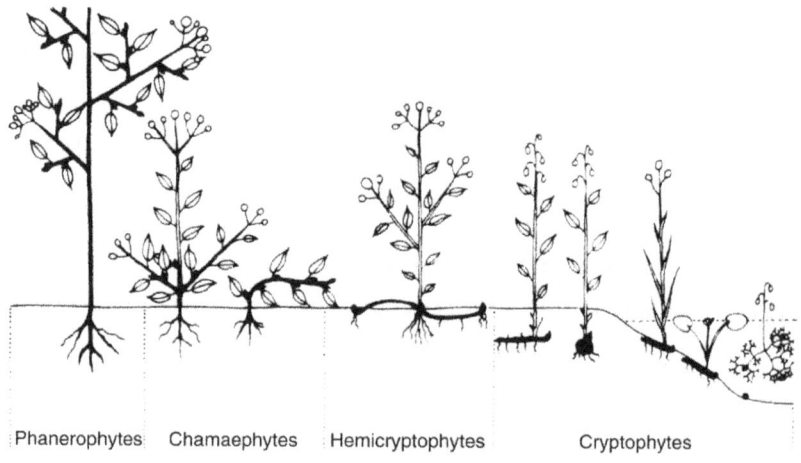

Phanerophytes   Chamaephytes   Hemicryptophytes   Cryptophytes

FIGURE 6.1 Plants can be divided into functional types based upon the location of the meristem that survives the most extreme filter in the habitat, which is usually cold. At the far left, phanerophytes (trees) expose the meristems on the tips of branches, while at the right, cryptophytes protect the meristems beneath the soil. (From Goldsmith and Harrison 1976, redrawn from Raunkiaer 1908.)

In practice, it is often useful to expand these five categories to include three (or four!) different kinds of phanerophytes and three kinds of cryptophytes, as well as adding stem succulents and epiphytes. This conveniently produces 10 categories, which is the system Raunkiaer used in many of his analyses comparing plants in different environments (recall Table 1.1).

Humboldt and Raunkiaer are rather old examples. We have kept them here because they remind us that some of the questions we are asking in ecology have deep roots. Both systems are based, however, mostly upon simple traits that are easily seen. As we have seen in Chapter 4, much better trait data are now available. So let us look at a few more recent examples using a wider array of traits. We will first look at herbaceous species in wetlands, and then woody species in tropical forests. Both of these are globally important ecosystems receiving intensive human impacts.

Functional Groups in Wetland Plants

Arnold van der Valk (1981) describes how plants in prairie wetlands respond to cycles of flooding. He used just three life history traits to designate 12 basic wetland "life history types." The three traits that he used were lifespan, propagule longevity and propagule establishment requirements. There were also three sub-categories for lifespan (annual, perennial and vegetative). The latter two categories distinguished species that are perennial but mostly short-lived from those perennial species that produce large and longer-lived clones. Propagule longevity also had two sub-categories: species with short-lived propagules that mostly disperse in space, and species with long-lived propagules that remain stored in the substrate. Finally, there were two sub-categories for establishment: those that establish on mud flats during low-water periods, and those that can establish in standing water. Hence, $3 \times 2 \times 2$ gives 12 different possible functional types. Flooding is then the environmental filter (or, in his terms, sieve) that determines which functional types are found in a wetland.

One can use many more traits to categorize marsh plants. In Chapter 4 (Table 4.3) we saw how Shipley et al. (1989) created a matrix containing 7 juvenile and 13 adult plant traits for 25 species of wetland plants. Juvenile traits included germination requirements and maximum relative growth rate. Adult traits were morphological attributes such as the distance between successive shoots and the form of the canopy. In juveniles, one of the most important traits was seed size, which was inversely correlated with germination rate in light. In adults, the most important axes of variation were the height of the plant and the width of the canopy. Perhaps the biggest surprise in the above work was the discovery that juvenile and adult traits were uncoupled. That is to say, the correlation matrices for adult traits showed no association with the correlation matrices for juvenile traits. Perhaps the traits required for regeneration in gaps are fundamentally different from the traits required to hold space as adults. This would mean that two categories, fugitive or stress-tolerant, could be constructed for each of two stages of life history, which produces just four functional types (Figure 6.2). The plants used in Shipley et al. (1989) occupy riverine marshes, which have intersecting gradients of flooding, disturbance and fertility. This makes them structurally more complicated than the prairie marshes studied by van der Valk. However, there is still a rather small number of functional types within this type of vegetation.

Continuing from Shipley et al. (1989), Boutin and Keddy (1993) created a larger matrix of 27 traits and 43 species in herbaceous wetland plants (Figure 6.3). In this classification, there could be as few as three basic functional types: ruderal, interstitial and matrix species. The matrix species consist of long-lived clonal plants like bulrushes and cattails that form much of the visually obvious vegetation. The interstitial species such as milkweeds and irises regenerate in various kinds of gaps within this matrix. The ruderals are different, and depend upon larger-scale disturbances, like drawdowns in water level that create mud flats. Ruderals often have large reserves of buried seed banks. Each of these groups contains two or three nested

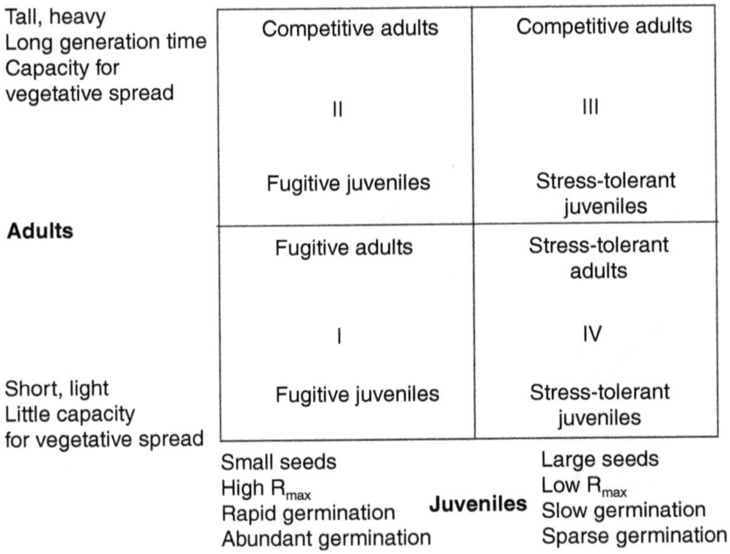

Tall, heavy Long generation time Capacity for vegetative spread	Competitive adults II Fugitive juveniles	Competitive adults III Stress-tolerant juveniles
Adults	Fugitive adults I Fugitive juveniles	Stress-tolerant adults IV Stress-tolerant juveniles
Short, light Little capacity for vegetative spread		

Small seeds High R_{max} Rapid germination Abundant germination   **Juveniles**   Large seeds Low R_{max} Slow germination Sparse germination

FIGURE 6.2 A multivariate examination of 20 plant traits in marsh plants produced four life history types. They appear to differ in tolerance to competition, disturbance and stress as juveniles or adults. (From Shipley et al. 1989.)

subtypes. For example, ruderals can be either "obligate annuals" or "facultative annuals." Both regenerate from buried seeds after a sudden change in the environment, like a sudden drawdown in water level, but the annuals live only one growing season, while the facultative annuals can reproduce the first year, but also persist for additional years, and may become a temporary presence in the newly established wetland vegetation. Interstitial species, by contrast, tend to persist in smaller gaps, where they disperse from one gap to another – that is, they disperse in space rather than time. A good example is Swamp Milkweed (*Asclepias incarnata*), which grows as a multi-stemmed tussock in small gaps, producing plumed seeds that drift to locate newly created gaps such as muskrat mounds or beaver lodges.

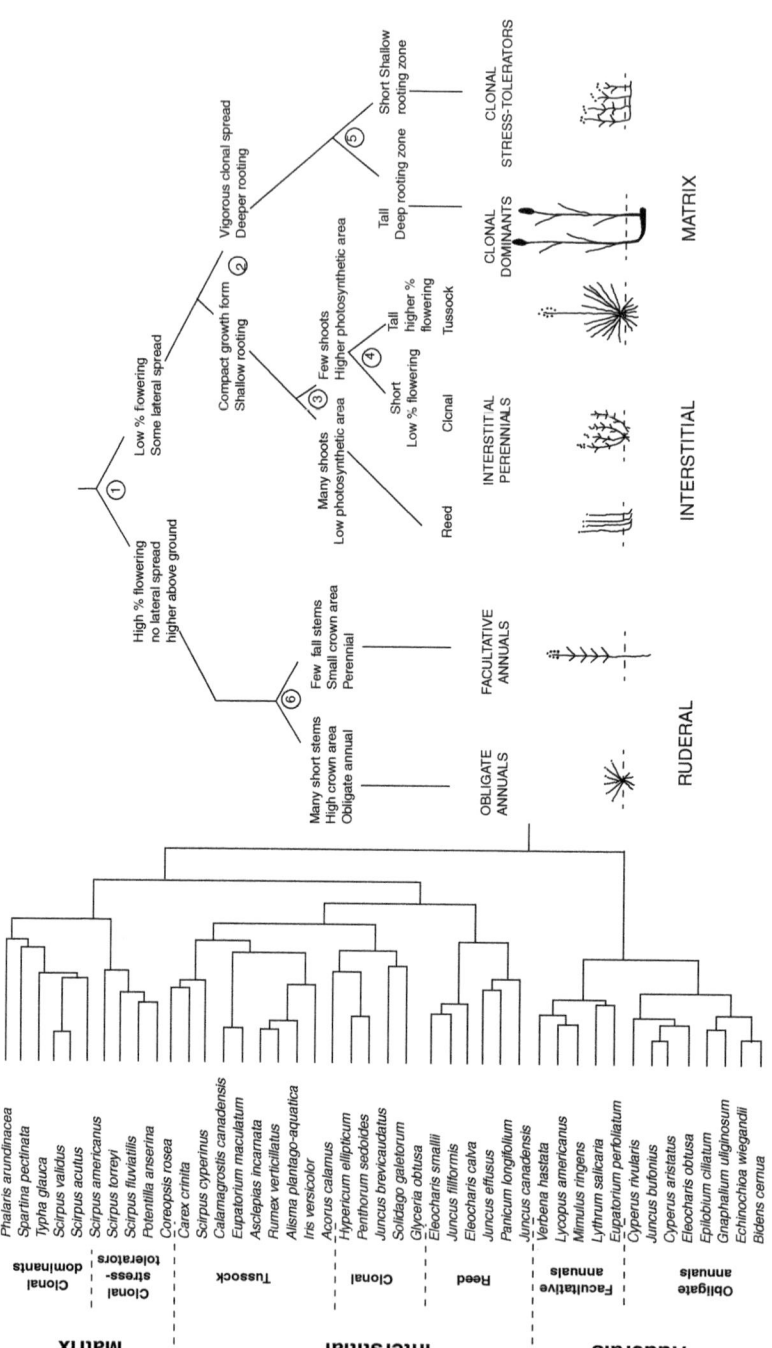

FIGURE 6.3 A hierarchical classification of functional types in herbaceous wetland plants. The left side shows the actual dendrogram for 43 species, while the right side offers an interpretation focused upon functional groups and traits. Note that there are either three or seven functional groups, depending upon the resolution required. (From Boutin and Keddy 1993.)

Functional Groups in Tropical Trees

Figures 6.2 and 6.3 include only herbaceous plants. Many landscapes are dominated by woody plants. In Chapter 5, we saw one example of a landscape with far more woody species, the garigue landscape of southern Spain (recall Figure 5.5), which is one example of the array of Mediterranean climate regions we discussed in Chapter 3. If we turn our attention to woody species, tropical forests provide the classic example, and those of the Amazon basin are notorious for their high species diversity (Hubbell and Foster 1986, Richards 1996). Early attempts at sorting these species into functional types (termed "synusiae") used mostly coarse morphological traits, recognizing groups such as trees, palms, climbers and epiphytes (Richards 1996). When size was used as a trait to delineate more functional types (e.g., height categories of ≤1 m, 1–3 m and ≥3 m; Whitmore et al. 1985), juveniles and adults of the same species could be assigned to different functional groups.

More recent work on tropical forests has used a broader array of traits, including leaf physiology and demographic characteristics. Here are three examples that have all been published just since we began work on this book.

Example 1: Tropical Rain Forest in Central America

In a tropical moist forest, Rüger et al. (2020) proposed that the functional types of tropical trees can be arrayed along two orthogonal trait axes. The first axis has long been used in tropical forest ecology; it recognizes the broad differences between fast-growing species that dominate early successional stages, and slow-growing shade-tolerant species that reach dominance at later successional stages. This distinction seems at first to assume that forest succession driven by natural disturbance regimes has been a prime factor shaping tree functional types (e.g., Connell 1978, Huston 1994). Given rates of natural disturbance from events like hurricanes, it probably makes more sense to think of these "early successional" functional types as being more the result of adaptation to ongoing natural disturbances. In this case, the species that are fast growing and light demanding might be considered equivalent to the interstitial species that occupy various kinds of gaps created by natural disturbance in herbaceous plant communities (Grubb 1986, Boutin and Keddy 1993). Therefore, the fast-growing and light-dependent species in tropical forests may be an integral component of old growth forest (Rüger et al. 2020). That is, they are not species that entirely disappear during succession, because frequent natural disturbances create the gaps in which they regenerate. More generally, such species have a specific kind of regeneration niche dependent upon gaps created by natural disturbance (Grubb 1977). There is an enormous number of natural disturbances that affect ecological communities: Keddy (2017, chapter 5) has given an overview of natural disturbances that affect ecological communities, ranging from the small scale (animal burrows, ice storms) to moderate scales (landslides, volcanoes) to the very largest scales (hurricanes, meteor impacts). It therefore makes sense that at least one group of species would be adapted to exploit conditions that arise after disturbance.

A second axis of variation in tropical trees may be more demographic in nature: a distinction between long-lived pioneers and short-lived breeders. Rüger et al. (2020) say that long-lived breeders "grow fast and live long, and hence attain a large stature, but exhibit low recruitment," whereas short-lived breeders "grow and survive poorly, and hence remain short-statured, but produce large numbers of offspring" (p. 165). They suggest that models of forest dynamics that incorporate these four groups (plus a fifth "intermediate" group) adequately describe expected changes in forest composition over several centuries.

Are these four (five?) groups (and their traits) sufficient for describing functional types in tropical forest? This takes us back to the question of how many traits we need to create useful functional groups (recall Table 4.1 and Figure 4.7). Rüger et al. (2020) note that their work does not include physiological mechanisms, and, as we have already described in this book, both leaf economics (Wright et al. 2004) and drought tolerance

(Choat et al. 2012) are important traits in trees. Other ecologists have also emphasized the importance of traits that confer resistance to herbivores (Coley 1983). Further, a focus solely upon tree traits omits other functional groups such as lianas and epiphytes, which are important components of diversity in tropical forests. The next example therefore uses physiological traits and a wider array of species.

Example 2: Tropical Montane Forest in Southeast Asia

Another team has offered a functional type classification within the tropical montane forests found on Hainan Island in China (Fan et al. 2019). Their functional group classification was constructed from a matrix of five ecophysiological traits measured on 87 plant species. The traits were net photosynthetic capacity (A_{max}), maximum stomatal conductance (g_{max}), water use efficiency (WUE), transpiration rate (Trmmol) and specific leaf area (SLA). Thus, they focused upon the traits that involve the use of water, uptake of carbon dioxide and the investment in leaf construction (recall Figure 4.6; Wright et al. 2004). Each trait was measured on 2–6 individuals for each species, and on 3–5 leaves for each individual. The species were then assigned to groups with a hierarchical cluster analysis (Figure 6.4). The result was eight functional types. As with the preceding example, the functional types could be sorted along a succession/natural disturbance gradient. The dominant functional groups in the primary stages had the highest net photosynthetic capacity and the highest maximum stomatal conductance. In contrast, in late stages of succession, SLA and WUE were highest. Functional groups seven and eight, for example, were mainly composed of late successional tree species (*Dillenia pentagyna, Nephelium topengii*). Group 4 was mainly composed of evergreen trees and shrubs. In contrast, group three had mostly shrubs and tall grasses. Group 1 consisted of a single species, the invasive herb *Eupatorium odorata*. Thus, in contrast to the first example, this classification system consisted of a broader array of plant types, but the trees themselves fell mostly into four groups, differentiated by leaf characteristics.

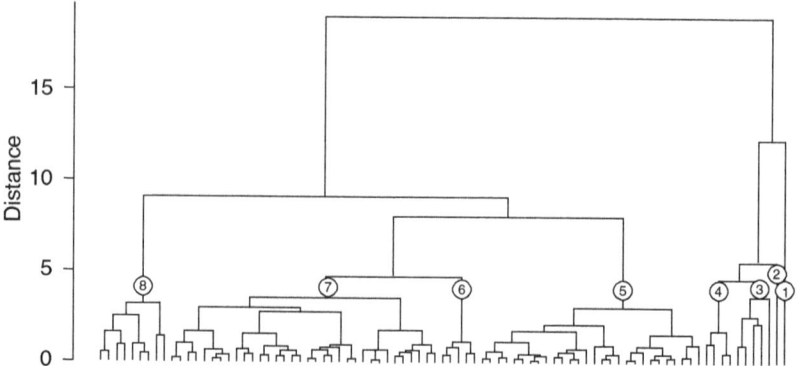

FIGURE 6.4 The dendrogram from a hierarchical cluster analysis of five physiological traits measured on the leaves of 87 plant species found in tropical montane forests of Southeast Asia. The clustering method used relative Euclidean distance with Ward linkage (adapted from Fan et al. 2019).

Example 3:   Tropical Rain Forest in the Amazon

The Amazon basin is a global reference point for tree diversity, and for problems with tropical deforestation. Fyllas and his team (2012) employed a trait matrix of 10 functional traits measured on 409 species, and also had environmental data from 53 forest plots. The trait data included leaf mass per area, branch xylem density and maximum tree height, among others. The investigation began with a PCA (principal component analysis) of the functional traits. This revealed three main axes. The first was a dimension of the leaf economics spectrum, reflecting leaves of different cost and durability. The second axis also reflected the leaf economics spectrum, in this case the investment of N and P in leaf construction. The third axis appeared to reflect whole-plant hydraulics and light acquisition, which seemed to be related to tree height. When these axes were related to the environmental factors in plots, the traits appeared to vary with soil characteristics such as fertility and texture. Fyllas et al. (2012) concluded that there were four principal groups, although it was also possible to make the case for six.

This foray into tropical forests illustrates a recurring theme in this chapter, and in this book. It illustrates a fundamentally important question in community ecology: how many traits and functional types are necessary to describe a particular community? Is there a correct or "natural" number of functional types, or are there only arbitrary positions within a divisive hierarchy that recognizes ever-finer distinctions among functional types? This second question is of course an old – even venerable – problem that arises when one is trying to find clusters of groups in multivariate data sets (Sokal and Sneath 1963, Gnanadesikan 1997). Statistical tools like Calinski pseudo-F (Caliński and Harabasz 1974) offer "stopping rules" to assist in making the decisions about just how many groups are present in a data set, but it seems that there is always an element of subjectivity. How many groups are enough, and how much detail is needed? This question remains embedded in modern statistics and computer science, where it is often described as the issue of "overfitting" (Christian and Griffiths 2016).

Returning to tropical forests, it would seem reasonable to divide the pool of hundreds of species of trees into subsets of different functional groups, but this leaves the thorny question of just how many such groups are enough, and, equally, which traits are most useful in recognizing these groups. As these three studies show, there is no agreement on which traits to use, or what the resulting functional groups are, although there appears to be an emerging consensus on the importance of traits that measure leaf economics and water use. There also appears to be an emerging consensus that the important environmental factors involve two environmental gradients: light (which is related to disturbance regimes) and soil fertility (which determines the availability of resources for construction of leaves and wood).

FUNCTIONAL GROUPS IN BIRDS

Birds do not have hands to process food, and so the food supply places strong selective pressure upon bill form, and provides a convenient means to sort species into basic feeding groups (recall Figure 4.3). At the finer scale, bills can vary in other attributes such as densities of comb-like lamellae used for filtering food particles from debris. Other attributes such as foraging habitat, nesting habitat and migration can be included (Weller 1999).

Many descriptions depend upon a keen eye for natural history rather than explicit data matrices. Here is an example of the latter: a regional study of 56 species of birds in Nicaragua (Pla et al. 2012). Traits included bill characteristics, wingspread, size, diet and migrant as opposed to resident status. Cluster analysis produced five functional groups (Figure 6.5): nectivorous, migratory generalist, insectivorous specialist, granivorous/omnivorous, and granivorous and carnivorous generalists (Figure 6.5).

More fine-scale analyses have also been done on the morphology and ecology of birds. One example was a study of 11 species found in winter forests in Iberia (Carrascal et al. 1990). In addition to bill traits, the study included another trait that is important in birds – the length of the tarsometatarsus, a bone found in the lower leg. In a PCA of these 11 "biometrical variables," size was found to be the most important factor, accounting for 43 percent of the variation in the data. Larger species had relatively shorter tarsometatarsi and tails. The second principal component was mainly associated with bill shape, and accounted for a further 26 percent of the variation. A second PCA was done on ecological data. Here, the first principal component distinguished between birds that fed on the trunk and thick branches, as opposed to birds that fed on twigs. The second principal component mostly distinguished between deciduous as opposed to coniferous trees. The morphological and ecological data show a relationship between the two types of data: birds exploiting tree distal parts and foliage are generally small and have a relatively long metatarsus. To put such work in the context of community assembly, note that all of these species, however, belonged to a single functional type ("guild"): pariform insectivorous birds found in winter forests of Iberia, such as the Coal Tit (*Parus ater*) and Short-Toed Tree Creeper (*Certhia brachydactyla*). Here is where

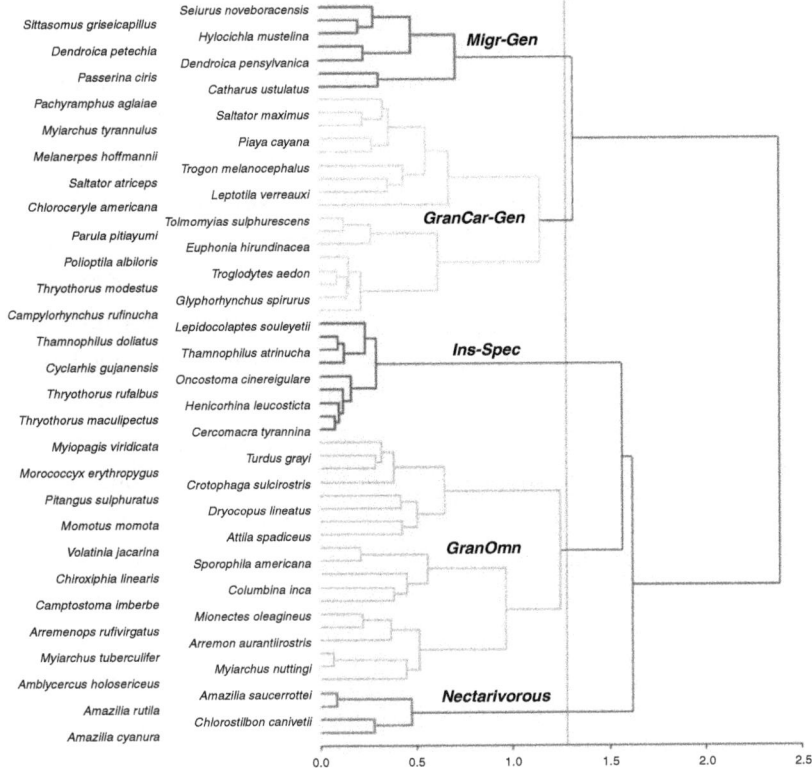

FIGURE 6.5 A set of bird species in Nicaragua classified into five functional types: insectivorous specialists (Ins-Spec), granivorous and carnivorous generalists (GranCar-Gen), granivorous and omnivorous (GranOmn), migratory generalists (Migr-Gen) and nectivorous. The clustering used the Gower similarity coefficient and Ward linkage. (From Pla et al. 2012.)

the issue of scale becomes important, as these data mostly describe variation within a functional group.

FUNCTIONAL GROUPS IN INSECTS

Aquatic invertebrates have been sorted into distinct functional groups, again based primarily upon feeding ecology, which in turn is determined in part by morphology (Cummins 1973, Cummins and Klug 1979). There are two principal sources of food exploited by stream invertebrates, detritus from the surrounding floodplain and plant material produced within the stream, including both periphytic algae and vascular plants. Functional types are frequently identified by the way in which they exploit these food sources. According to Cummins and Klug there are six groups: shredders, collectors, scrapers, macrophyte piercers, predators and parasites (Figure 6.6). Habitat, dispersal, life cycle and size can be used to expand the system. Since there may be hundreds of taxa living in a stream, this provides a rapid method for sorting them into ecologically meaningful groups. As just one example, a single study in the Mara River basin in Kenya found 20,757 individuals from 109 taxa in just 20 sites (Masese et al. 2014). In most such studies, the number of "taxa" significantly underreports the true biological diversity, since taxon categories may be only at the generic level, or may use morphospecies. The most common species were crabs (*Potamonautes* spp.) and crane flies (*Tipula* spp.).

An alternative method for classifying stream invertebrates into feeding groups is to look not directly at their feeding morphology, but instead at their diet, by directly examining their gut contents. Continuing with the example from the Mara River basin in Kenya, the contents of the guts were sorted into different food types (vascular plant material, large particles, small particles (<50 µm), algae, animal material and inorganic material such as sand or silt). The taxa were then sorted into functional groups using cluster analysis (Figure 6.7). Six functional types could be distinguished.

In the study of stream ecology, the recognition of functional feeding types is not considered merely a tool for description. The relative abundance of different feeding types is considered a way to measure the amount of pollution rivers are receiving, for example, as well as providing information on the balance between autotrophy and heterotrophy (Cummins et al. 2005).

FUNCTIONAL TYPES OF FISH

Many small, northern lakes have rather low numbers of fish species, and most ecological studies are done on a species basis. For example,

Functional group	Dominant food	Feeding mechanism	Order example	
Shredders	Living tissue	Herbivore	Lepidoptera	
	Decomposing tissue	Detritivore	Plecoptera	
	Wood	Gouger	Coleoptera	
Collectors	Decomposing organic matter	Detritivore	Collembola	
Scrapers	Periphyton	Herbivore	Coleoptera	
Macrophyte piercers	Living tissue	Herbivore	Neuroptera	
Predators	Living tissue	Engulfer	Megaloptera	
		Piercer	Hemiptera	
Parasites	Living tissue	Internal and external parasite	Hymenoptera	

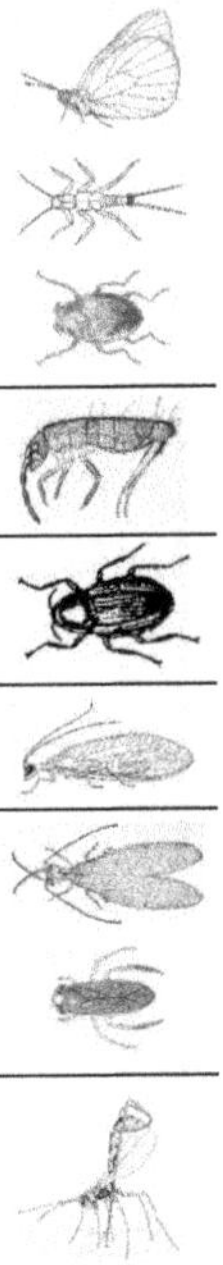

FIGURE 6.6 The insects in wetlands can be divided into six functional groups based upon their dominant feeding mechanism. (From Keddy 2010, modified in part from Merritt and Cummins 1984.)

limnologists have studied the assembly of fish communities in Wisconsin and Michigan, where there are thousands of small lakes and a small number of species of fish (Tonn and Magnuson 1982, Magnuson et al. 1989). Magnuson and his workers concluded that these lakes can be sorted into just a few distinctive community types. The most important filters were low oxygen levels in winter and low pH, both of which tended to remove large predatory fish. In lakes with predators, the composition differed between the two main predators: bass or pike. Thus, we can indeed have community assembly rules based upon species names. However, these lakes have a relatively small number of fish species. As the number of

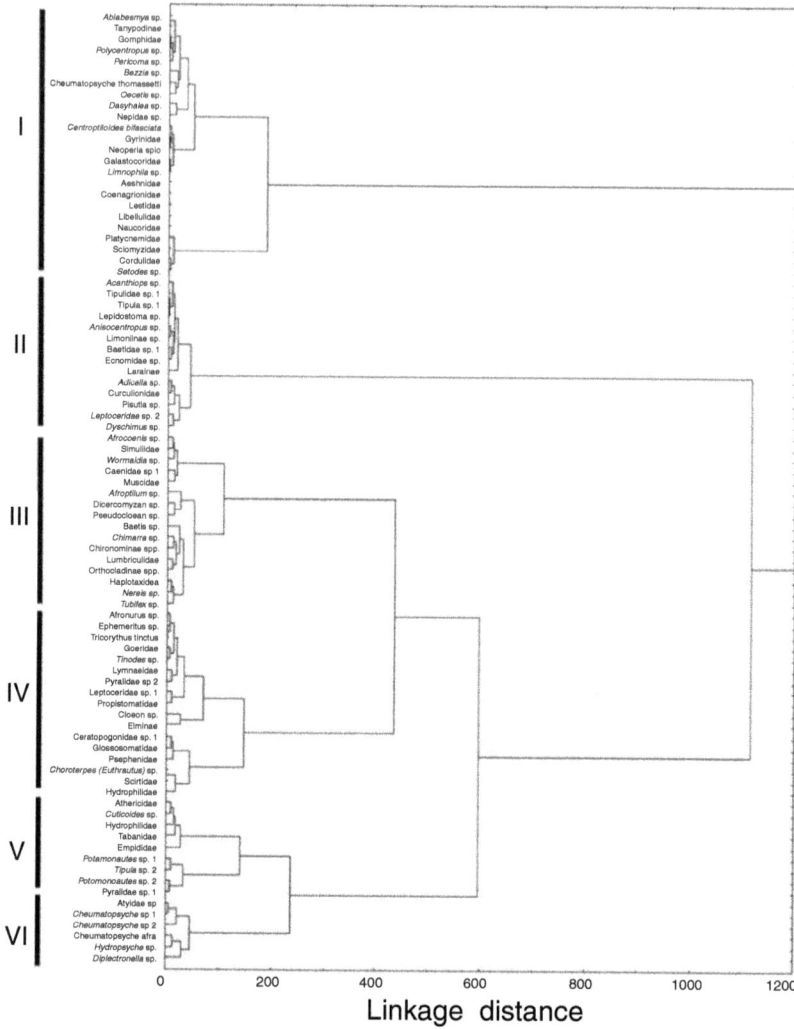

FIGURE 6.7 Invertebrate taxa in Kenyan rivers were clustered into functional feeding groups based upon the proportion of different food types in their guts. Group I consists of predators, group II of specialist shredders, group III of mostly collectors, group IV of mostly scrapers, group V of predators and generalist shredders and group VI of generalist collectors. (From Masese et al. 2014.)

species increases, or as the scope of the study expands to taxa beyond fish, we quickly reach the point described by Rigler: too many species.

One simple way of classifying fish by functional types is to focus upon diet. Wootton (1990) divides fish into four main feeding groups: detritivores, scavengers, herbivores and carnivores. The carnivores, in turn, are divided into four main groups, again based upon diet: benthivores, zooplanktivores, aerial feeders and piscivores. This gives us seven functional feeding groups.

When you move to the tropics, the sheer diversity of fish species in large rivers like the Congo, Mekong or Amazon makes some sort of functional classification essential. Consider the Amazon (Figure 6.8). The majority of the described species are characoids and siluroids. The characoids are mostly laterally compressed, silvery, open-water fish that are active by day. They have undergone spectacular adaptive radiation, include both the fruit-eating *Colossoma* and carnivorous piranha, and are probably "one of the most diverse groups of living vertebrates" (Lowe-McConnell 1975, p. 38). The siluroids ("catfish"), by contrast, are mostly bottom-living and nocturnal. They include piscivores, planktivores and even parasites. Apart from the characoids and siluroids, another notable group is the gymnotoids, the nocturnal electrogenic fish. There are also a substantial number of fruit-eating characoids, which live in flooded forests and disperse tree seeds (Goulding 1980). Hence, the system with seven functional types based upon diet may be too simplistic.

How have tropical fish ecologists approached this problem? There are quantitative functional type descriptions for selective habitats, such as riffles (Casatti and Castro 2006), floodplain lakes (Siqueira-Souza et al. 2017), forest streams (Brejão et al. 2013) and estuaries (Pessanha et al. 2015). In the latter case, fish were collected in the Rio Mamanguape estuary in northeastern Brazil. There were 11 morphological attributes measured, including total length, body height, body width, mouth width, and mouth height. These data were then used to classify the fish into groups. The fish were also independently divided into six trophic groups based upon their diet: detritivores, benthivores, zooplanktivores, insectivores, macrocarnivores and omnivores. There were thus two methods used to assign

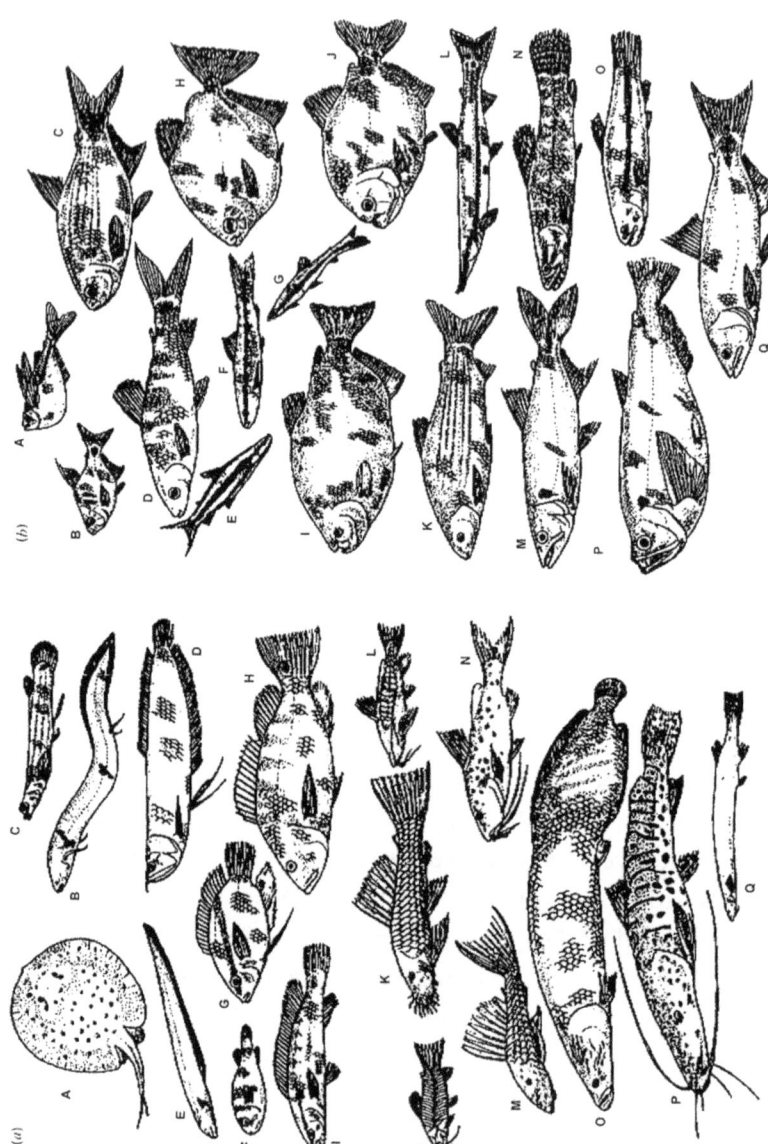

FIGURE 6.8 Some representative siluroid (a) and characoid (b) fish from the Amazon River system. [From Lowe-McConnell 1975.]

species to functional groups, and the authors conclude that ecomorphology is indeed associated with diet.

Figure 6.9 was based on morphological traits measured on caught fish. Instead, one could focus on the feeding behaviour and location of live fish. Brejão et al. (2013) observed the feeding behaviour of 73 species of Amazonian fish and used these observations to sort them into "functional trophic groups." The classification started with 15 predefined feeding tactics: surface pickers, drift feeders, roving predators, stalking predators, ambush predators, mud-eaters, diggers, grubbers, nibblers, sit-and-wait predators, crepuscular predators,

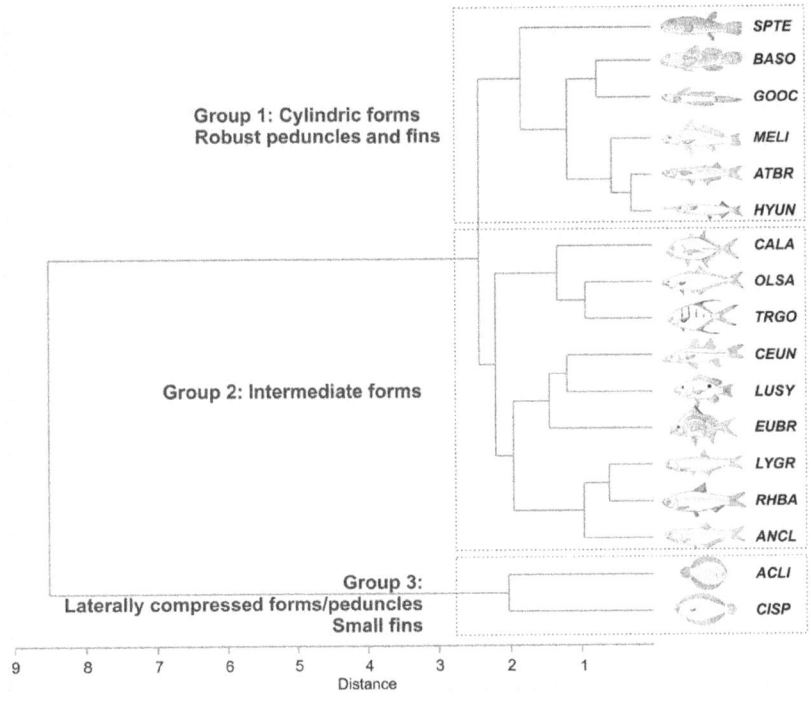

FIGURE 6.9 A classification of functional types in 17 species of tropical estuarine fish in northeastern Brazil based upon 11 morphological traits. (From Pessanha et al. 2015.)

nocturnal predators, grazers, parasites and invertebrate pickers. These were then modified by feeding location to produce a total of 18 groups (Figure 6.10). For example, "browsers" (at the top of the right-hand column of the figure) are nektonic species that bite off small pieces of macrophytes, while the group "diurnal channel drift feeders" consists of "nektonic species that collect food items drifting at mid-water and at the surface."

To put this work into a global context, let us close with the functional diversity in fish for each of the world's six terrestrial biogeographic realms, using measured traits on fish morphology (Toussaint et al. 2016). For each species, there were 11 morphological measurements that were used to calculate 9 unitless ratios describing the morphology of the fish head, body, pectoral and caudal fins. Size was a tenth trait. These traits were thought to describe "species strategies for food acquisition and locomotion." The authors observed that many other important traits, such as gut length or physiological traits, could not be included because of the absence of data for many species. (This is of course why Humboldt and Raunkiaer also had to depend upon morphologically obvious traits for their global-scale descriptions.) These data were used to create a five-dimensional morphological space. The Neotropics had by far the largest functional diversity. Although the Neotropics comprise less than 15 percent of the world's continental surface, they have 75 percent of the world's functional diversity in fish. Much of this can be explained by the uniqueness of certain Neotropical functional groups, particularly those that are elongated surface predators or flattened benthic algae browsers.

With these new examples, let us reconsider our opening question as to whether it will be possible to predict the number of functional groups *a priori* from foundational principles. For the fish in the Congo River, Wootton's system would give seven functional feeding groups. And, if we had tested our assumptions using the Mekong River data, we might have found the system sufficient for many purposes. But neither of these would have prepared us for the Amazon. The same functional groups certainly exist, but we

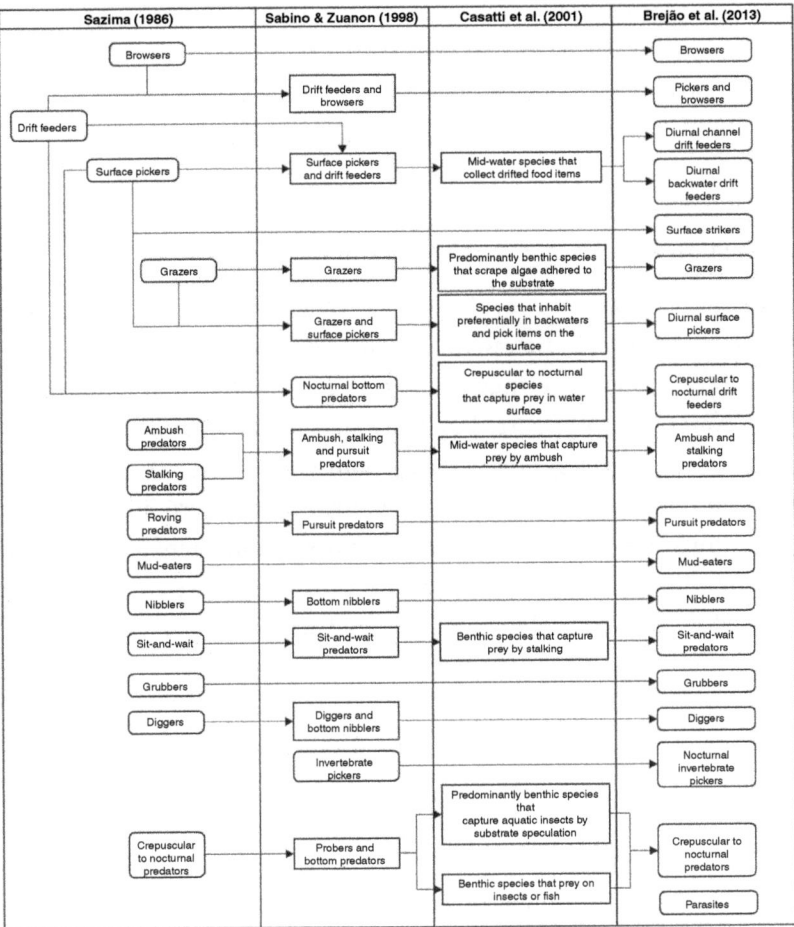

FIGURE 6.10 Functional feeding classifications of freshwater fish in the Amazon River compared across four studies. (From Brejão et al. 2013.)

might not have had grounds to expect so many fish that feed on tree fruits. And as for the entirely novel fish forms, who would have expected to find

> unique functional attributes such as i) extremely elongated fish
> with a large terminal mouth and a high caudal peduncle throttling,
> corresponding to mobile surface predators such as some of the

Beloniforms, or ii) dorso-ventraly flattened fishes, with a ventral mouth located below the head and a small caudal peduncle throttling, mainly corresponding to benthic algae browsers with limited swimming efficiency such as some Loricariid species.

(*Toussaint et al, 2016, pp. 3–4*)

FUNCTIONAL TYPES OF MAMMALS

Now consider mammals, which can be divided into functional groups based upon body size, diet (which can often be inferred from dentition) and habitat type. Severinghaus (1981) proposed that mammals in North America can be sorted by body size (small, medium and large) and diet (carnivorous, herbivorous, omnivorous) for a total of nine functional groups ("guilds" in his words). Since there are no small omnivorous mammals recognized in his system, there are actually only eight functional groups (Figure 6.11). Each of these can be further subdivided. For example, medium-sized carnivores are divided by feeding method: digging (e.g., badger), semi-aquatic (e.g., otter), tunnel searching (e.g., weasels), arboreal (e.g., marten) and stalking (e.g., raccoon). At the finest scale, there are 30 types of mammals in North America.

FUNCTIONAL TYPES OF FUNGI

Field identification of macroscopic fungi often begins with knowledge of morphological traits of their reproductive phase, using visually obvious characters such as size, shape, colour and gross morphology of the "mushroom," as well as whether spores disperse via pores or gills. To divide fungi into functional groups, however, one mostly requires quite a different kind of information – the resource use by the hyphae themselves, a trait that is not obvious on inspection. One approach recognizes three categories of such functional traits (Treseder and Lennon 2015). The first set of traits deals directly with decomposition activity, whether the fungus can degrade cellulose or lignin. A second set of traits is involved in

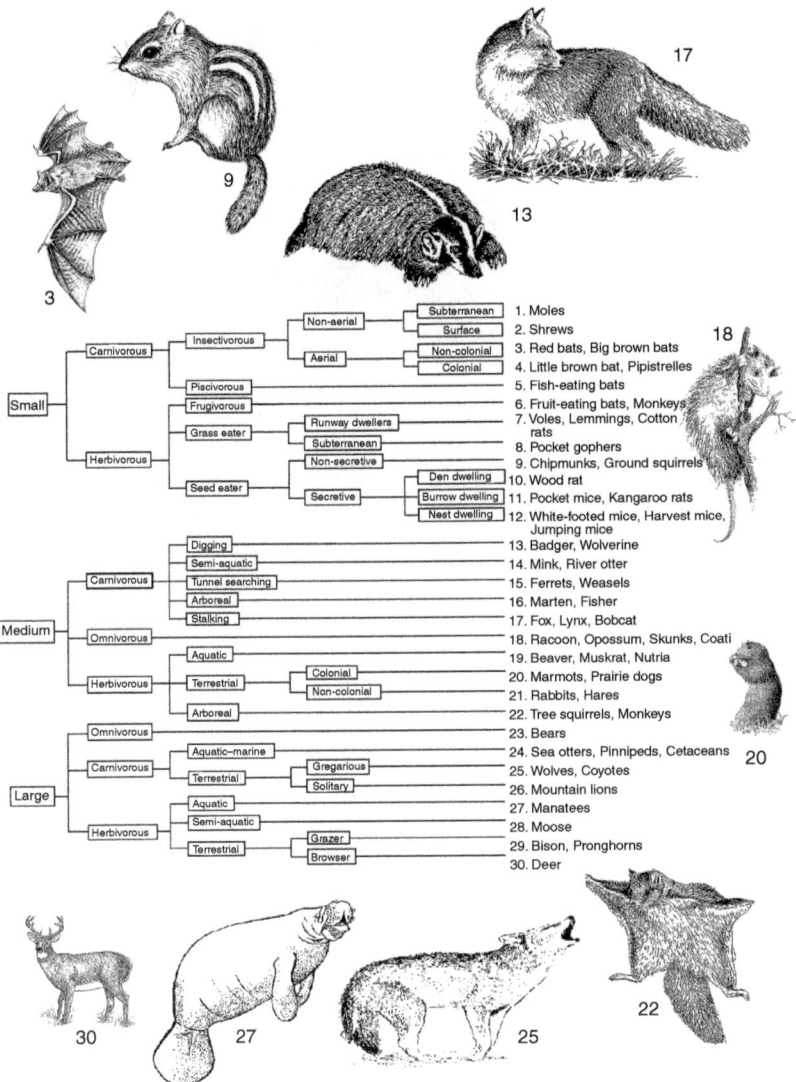

FIGURE 6.11 A functional classification for mammals of temperate regions based upon non-marine mammals inhabiting the continental USA. The number of functional types ranges from 8 to 30 depending upon the resolution required. (From Severinghaus 1981; sketches by R. Savannah, US Fish and Wildlife Service.)

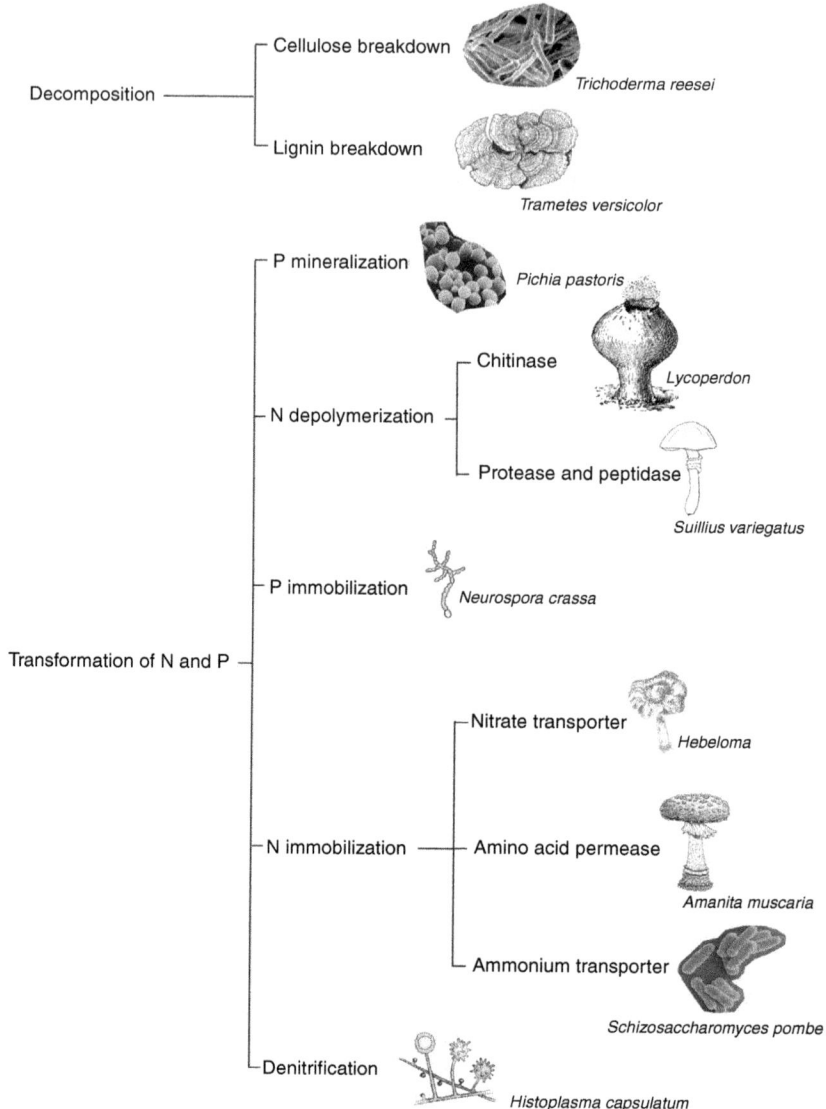

FIGURE 6.12 A functional classification of fungi based on traits that directly influence ecosystem processes. (Prepared using information in Treseder and Lennon 2015.)

nitrogen and phosphorus transformation. A third set of traits is associated with stress tolerance, that is, properties that allow certain fungi to remain active in dry, hot or cold conditions. Thus, fungi can be sorted into functional groups based upon their physiology (Figure 6.12). The genetic database for such traits is expanding with the 1000 Fungal Genomes Project (http://1000.fungalgenomes.org/home/). There are many more traits that might be used for functional classification (Table 6.1).

SOME METHODOLOGICAL ISSUES

In trying to think through the actual role of functional groups in community assembly, it is easy to be distracted by issues of methodology. There are at least five issues that need to be considered in future work on functional groups.

Table 6.1 *Some traits of mycorrhizal fungi that might be used for functional classification (adapted from Allen et al. 2003, table 2)*

1. Specificity of association with host
2. Biomass and morphological/structural traits
   Size and spatial distribution of individual fungal genets
   Total biomass per soil volume
   Intensity of root colonization
   Structure and thickness of EM fungal mantle
   Presence and abundance of AM arbuscules
   Mycelium architecture, hyphal branching and
      hydrophilic or hydrophobic properties of mycelium
3. Other life history traits
   Lifespan of individual fungal genets
   Longevity and turnover rates of mycorrhizal roots
   Primary strategy of root colonization: spores vs. mycelial inoculum
4. Physiological/biochemical characteristics
   Ability to utilize organic sources of nutrients
   Ability to mobilize nutrients from minerals (rocks)
   Ability to excrete organic acids

Table 6.1 (*cont.*)

Membrane transporters present in hyphae
Nutrient uptake efficiency
Nutrient concentration of fungal tissue
Production of antibiotics, phenolics or other secondary metabolism
 compounds
5. Other functional traits
Optima and tolerance range for abiotic factors
Competitive ability against other mycorrhizal fungi
Competitive ability against saprotrophic fungi or soil bacteria
Responses to other environmental factors

Scale. Is the classification meant to include all the groups of organisms found in the landscape, or merely a subset? For purposes of community assembly, it is certainly preferable to have the entire species pool classified (e.g., Figure 6.10), but a majority of studies pick only a subset of habitats or species. Hence, they are incomplete, and not terribly useful except for narrowly defined purposes.

Resolution. How many groups is enough? We saw with wetland plants that there were as few as three functional types, with fish as few as seven, and with mammals as few as eight. These categories may be too coarse for many purposes. In the nested classification of wetland plants (Figure 6.3), three groups can be expanded to seven. In the nested classification system for mammals, eight groups can be expanded to as many as 30. Which leaves as an open question, just how many categories are necessary to deal with tropical trees or tropical freshwater fish?

Data sources. Some studies use mostly qualitative data, such as the bird bills in Figure 4.3 and the mammal functional groups in Figure 6.11. Other studies use quantitative data, but focus on morphological measurements, such as the fish functional (morphological?) groups in Figure 6.8. It is necessary for ecologists to move beyond morphological traits to at least include physiological traits, or, better still, comparative traits determined by large-scale screening (recall Chapter 4). These kinds of data were included in the studies that yielded Figures 6.2 and 6.3. Such studies are still relatively uncommon, although growing number, as illustrated by the inclusion of ecophysiological traits in delineating functional types of tropical trees (Fan et al. 2019).

Still, we are left with a serious problem: if existing models that use functional groups do not provide useful predictions for community ecology, we cannot be certain of the cause: is our basic approach to community ecology wrong, are the models themselves faulty, or is it simply that the functional groups are still too poorly constructed to be useful?

Methodology. Methodology is to some extent determined by the type of data available. Even when quantitative data are available, there is still an array of choices to be made in classification procedures, such as the choice of similarity measures and the choice of clustering algorithm (e.g., Sokal and Sneath 1963). The issue of how many groups to recognize is not solved by these quantitative techniques, although certain tools can help reduce some of the subjectivity (Caliński and Harabasz 1974). For the near future, it would seem that there will always be an element of subjectivity in such analyses.

Purpose. It is apparent from reading many examples that sometimes people seem to make functional groups (particularly ecomorphological types) just because they have some easily measured trait data and a statistical package. This is not unlike the style of community ecology in which one collects vegetation data in quadrats and then subjects it to ordination. What exactly is the purpose? If the purpose is to uncover the fundamental units from which natural communities are assembled, then why would you pick small subsets of species by habitat type (e.g., riffle in a river) or taxonomic category (pariform insectivorous birds) and only then look for ecological and or morphological patterns within this narrow subset? The subset has already lost the important information of differences among groups in habitats and traits. That is, in such studies, it often seems that the researcher first excludes ecologically meaningful variation, and then looks for patterns within the residual variance! Perhaps this is why such studies make many references to coexistence, as if the phenomenon of coexistence grants permission for endless studies of local trait variation in selected groups of organisms. Such studies could go on for generations – and just might do so. Perhaps they illustrate yet another rebirth of Zeno's tortoise. We suggest that it is better to look at larger-scale patterns: there is nothing wrong with choosing a good-quality data set, with ecologically meaningful traits, and simply providing a reliable and objective functional classification for use by other scientists.

ON THE REALITY OF FUNCTIONAL GROUPS

These examples raise another thorny problem: the reality of such groups altogether. Here we need some caution, since we have historical examples of arguments about whether nature is continuous or organized into groups (one way of describing the Gleason vs. Clements debate). We don't want to lead ourselves into a divisive argument over two stereotypical views of reality, yet we must admit that there is a potential problem, and one that we are not going to magically resolve in this chapter, or this book. Are functional groups real or imaginary? The truth probably is that they lie uncomfortably in between two ideological extremes. The same by the way, seems to be true of the Gleason vs. Clements debate. Yes, nature is non-random, and there is recurring pattern in plant communities in gradients, but no, the communities are not clear-cut with hard boundaries. You can read the details in chapter 11 of *Plant Ecology* (Keddy 2017).

From one perspective, creating functional groups is an arbitrary process with the simple task of creating groups of similar species, mostly for the ease of human thinking. From the other perspective, it is a matter of uncovering real natural groups that represent certain combinations of traits that recur in nature, presumably because there are certain regions of the underlying ecomorphological landscape, a sort of hidden structure that guides organisms into converging life forms.

A traditional approach to supposedly address this distinction is to collect data and then subject them to a multivariate cluster analysis and then let the data decide if there are groups or not. Here again the outcome is more or less given. The multivariate techniques that exist to find clusters will, when given a matrix of species and attributes, indeed find clusters. In the same way, if exactly the same data are subjected to a multivariate technique to find gradients in traits, gradients will indeed emerge. Back to mammals: is it useful to divide them into three size groups, or should we accept that there is actually a gradient of body sizes and work directly with measures of body mass? For many years, plant ecologists argued about whether communities occurred as clusters or gradients, without apparently being willing to concede that the answer mostly depended, *a priori*, upon the multivariate technique that was applied. Of course, they were thinking about communities rather than functional types, but the underlying issue is identical: is nature in some way categorical, or just continuous?

Then there is a deeper philosophical issue. What is real and what is not? We would have to pay homage to both Plato and Descartes. The latter insisted on his own existence (I think, therefore I am), while the Buddha argued that even the concept of "I" is an illusion. This is why books or papers that pose questions like "are communities real?" or "are functional groups real?" are not only misleading, but annoying. They divert attention and waste intellectual effort. To avoid these digressions, we will simply note here that functional types provide a useful way of thinking about the natural world. They are a tool. We do not get into discussions about whether uranium atoms are real, or whether chickadees are real. Both have certain measurable properties, and it is convenient to use the words uranium and chickadee in science. The same is true of functional groups: we construct them and use them, either for describing wild nature, or, better still, for making predictions about wild nature.

THE UNDERLYING CAUSES OF FUNCTIONAL GROUPS

We now address two oversights in this chapter through a digression into theoretical morphology, fitness landscapes and natural selection. First, we have not yet said much about the evolution of functional groups. Second, we have not mentioned molluscs. There is a reason we have mostly side-stepped evolution: as we explained in Chapters 1 and 3, the process of community assembly occurs at rather short time scales, allowing us to assume that the process of evolution is practically fixed (recall Figure 1.5). Yes, evolution is the causal factor that determines both the number of species that exist and the traits they possess. But the time scale of community assembly is rapid in contrast to the much slower evolution of traits and species. As a crude analogy, when we heat water to make our morning cup of coffee, we do not have to worry about the supernovas that created the elements occurring in the water and the coffee. Nor, on a shorter time scale still, do we need to consider which rainstorms deposited that water, or which plantation grew those coffee beans. (Unless we are using very specially sourced bird-friendly coffee beans, and even they now come from multiple plantations.) So, the question of how many functional groups are found in a community requires an answer at two quite different time scales. At the community scale, there is the question of how many functional types will fit into a community. Are there some rules? That remains our focus in this book. But it is necessary to at least introduce the evolutionary context, where the broader question is, and how many kinds of functional types will evolve in response to a set of environmental filters. This question is surely related to how many trait axes are necessary to fully describe the n-dimensional trait space a group of species will occupy.

The question of just how many axes are enough has its own lineage in articles dealing with evolutionary ecology. This now also brings us to molluscs. In 1966, David Raup wrote about the minimum number of axes necessary to describe all the different kinds of shells of marine invertebrates. He called this type of work "theoretical morphology," and settled on three axes as being sufficient. Even just three axes produced a volume containing a wide array of possible shapes and sizes. Remarkably, although there are a great many sizes and shapes of molluscs, the observed volume of shapes is vastly smaller than the theoretically possible array of mollusc shapes contained within this three-dimensional space. To repeat, the actual ecomorphological space is only a small fraction of the theoretical ecomorphological space. Interestingly, a similar conclusion was reached by Díaz et al. (2016) about the shape of trait space occupied by vascular plants – even though six traits were measured, the trait space was remarkably flat and disc-like. This led Raup to conclude that certain shell forms "are theoretically possible but do not occur in nature" (p. 1178). The reasons why this should be the case are complicated. Perhaps certain shapes are physically impossible. Perhaps certain shapes do not correspond in any useful way to the actual habitats or resources found in nature. Perhaps natural selection simply cannot produce certain shell shapes.

The divergence between observed and theoretical forms within n-dimensional space is a complicated problem. It likely means that all of the functional groups described in this chapter are dimensionally much smaller than the theoretical possibilities. In general, this situation requires us to consider the constraints on natural selection, and particularly the concept of the *fitness landscape*, introduced by Wright (1932), which is now considered to have a "central and special position in evolutionary theory" (Svensson 2016). These constraints on the traits and forms of living organisms have been more popularly discussed by both Dawkins (1996) and Lane (2010).

Our point is that evolutionary biologists have thought carefully about how many axes are needed, and what proportion of the potential space is actually occupied by existing living creatures, and how these are related to fitness landscapes. Continuing with molluscs, and marine invertebrates, and axes in n-dimensional space, and functional groups, it is noteworthy that another scientist has since extended Raup's list from three axes to ten (Tursch 1997)! In this chapter, we have tried to take an empirical approach, mostly describing which kinds of traits produce which kinds of functional groups. From this perspective, the fundamental questions remain: (1) How many axes do we need to measure? (2) Which traits best represent these axes? (3) How many functional groups are typically found in different communities? Deeper questions of how and why remain to be explored elsewhere, as illustrated by both Raup and Dawkins.

FUNCTIONAL GROUPS, TRAIT MATRICES AND ECOLOGICAL ASSEMBLY

Overall, it is apparent that for most habitats, even really important ones like tropical forests, we lack functional group classifications that are based upon quantitative traits, particularly classifications that use ecological or physiological traits from comparative studies. Overall, only a small proportion of functional group studies use quantitative data on ecological traits, and only a small proportion of these use multivariate clustering methods. Perhaps these quantitative procedures are not necessary in all situations, and perhaps natural history knowledge alone is suitable for many purposes, but it does raise the question as to whether it will be possible to use qualitative classifications of functional types to build quantitative models of ecological communities.

   This likely means that it will be necessary to invest time and effort into building such classifications. There is no need to try to justify functional group studies by claiming that they (might) "provide insight" into coexistence. Nor is there a pressing need for more studies that pick a narrow habitat or small group of species and explore variation within them – this is the ecological equivalence of removing main effects from an analysis of variance and then trying to look for patterns in the residual data. Or, to an architect thinking about how to build a house, but starting with a careful multivariate analysis of the doorknobs that are available for the front entrance. Or, to return to furniture assembly, it is like exploring the multivariate patterns in screws or nails, rather than on the assembly of the bookcase itself.

An alternative strategy for community ecology is to try to bypass functional groups entirely by using quantitative data in trait matrices. We will explore this prospect in the next chapter. Again, however, as we saw in Chapter 4, often the necessary trait data are scarce. Thus, whether or not we use functional groups for community assembly, we still need good-quality trait matrices. Otherwise, people seem to keep falling back upon simple morphological traits because they cannot find anything better, which takes us into endless cycles of trying to build predictive models from inadequate trait data.

It is not yet possible to say whether it will be necessary, or helpful, to have functional group classifications as a tool for the assembly of ecological communities. It is probably premature to make this decision, given the inadequacy of most of the functional group classifications that are available. However, since functional group classifications are useful for a wide range of descriptive purposes and for habitat monitoring, it seems like a task that deserves more serious effort. Enormous effort has gone into constructing phylogenetic classifications systems for life forms, and if even a small fraction of this effort were put into functional classifications, we would be more knowledgeable about the diversity of life forms in general.

There do not seem to be any *a priori* grounds for predicting how many functional groups a habitat can support. This may be a property that can be determined only by describing the functional groups we actually observe in each type of habitat. While it is true that convergent evolution in response to filters might be expected to produce just one ecological response, we have good examples where this is not the case. We could not predict *a priori* that there would be at least four functional types for plants growing in a quite simple habitat – shallow water. Indeed, there are other classifications that yield even more groups of shallow-water plants (nine in Figure 6.13). In the same way, it is unlikely that we would have any *a priori* grounds for predicting that Neotropical rivers would have more life form variation

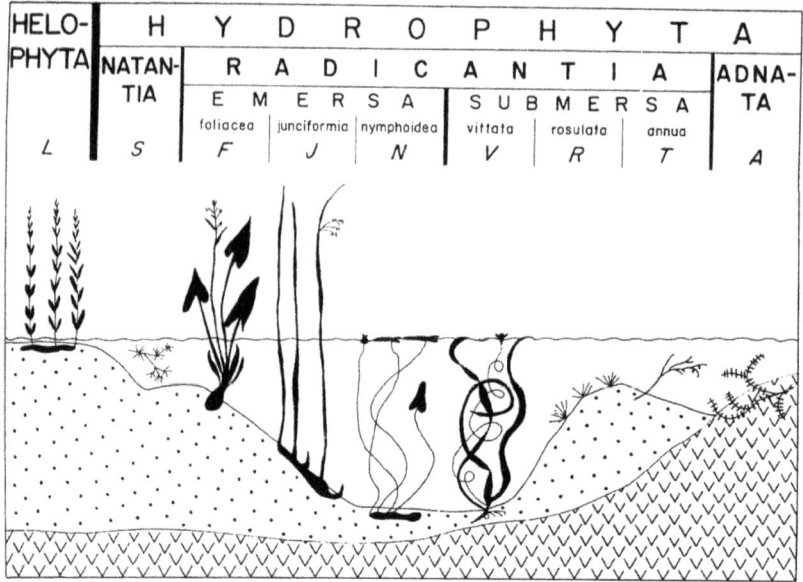

HELO-PHYTA	H Y D R O P H Y T A								
	NATAN-TIA	RADICANTIA							ADNA-TA
		EMERSA			SUBMERSA			
		foliacea	junciformia	nymphoidea	vittata	rosulata	annua	
L	S	F	J	N	V	R	T	A

FIGURE 6.13 Shallow-water plants may be classified into nine different functional groups. (From Dansereau 1959.)

that tropical rivers in Africa and Asia. This does not mean that we cannot tabulate such data for future use. For example, it may be possible to determine that each type of environment has a prescribed number of functional groups as a templet for this habitat. If so, then the problem of community assembly is simplified to predicting the relative abundance of each functional group. The question of how many similar species occur within a functional group is then conveniently seen to be a second and independent line of inquiry in community ecology.

KEY POINTS OF THE CHAPTER

- We suggest that to assemble an ecological community, it may be helpful to know not only how *many* parts are there, but what *kinds* of parts are there (e.g., in building a house, a doorknob cannot replace a window).

- Communities require at least two classification systems that provide simultaneous and somewhat contradictory lists of parts: phylogenetic (how many?) and functional (what kinds?).
- These two classification systems can be arranged hierarchically so that a large number of parts (species) are nested within a smaller number of groups (functional types).
- There are standard algorithms that can be applied to trait matrices to objectively classify species into similar groups. The better the quality of the data in the matrix, the better the resulting classification of functional types.
- Functional types can themselves be organized hierarchically into groups of increasing size and decreasing within-group similarity.
- Even with objective classification techniques, it is difficult to know how many groups exist, and the number selected may be somewhat arbitrary and depend in part upon the processing limitations of human brains.
- We suggest there may be no way to tell, *a priori*, how many functional types we can expect to find in a specified landscape or habitat.

7 Predictive Models of Community Assembly

Community Assembly by Trait Selection (CATS). An application to North African arid landscapes. How many traits are enough? Community assembly using Traitspace. An application to conifer forests in southwestern North America. Applying both methods to a kettlehole wetland.

PREDICTION IN COMMUNITY ECOLOGY

We have established that there are measurable relationships between traits and environments, generally in Chapter 4, and specifically using quantitative analyses presented in Chapter 5. What is next? Well, it is one thing to quantify relationships between traits and environments. The next step is surely prediction, which Peters (1992) regarded as one of the most important challenges for ecology. At the risk of being repetitive, we will restate that describing patterns in traits is certainly an improvement over simply describing patterns in species. But both are mere descriptions of patterns seen in nature.

Going all the way back to Figures 1.1 and 1.5, can we use trait–environment relationships to predict the occurrence of the species in a habitat? A successful model of community assembly should take our understanding of how traits interact with the environment, and then use it to predict how abundances of every species in the species pool change along environmental gradients (Shipley 2010). In Chapter 1, we said that the central challenge in community ecology is to produce vectors of species abundances from trait matrices. In other words, our challenge is "to say which grain will grow and which will not" (Banquo, from Shakespeare's *Macbeth*, Act I, Scene 3). Here, we will describe two models that do just that.

We will first describe the Community Assembly by Trait Selection (CATS) model (originally referred to as "maxent"), which uses a system of linear equations built with average trait values to predict species relative abundances (Shipley et al. 2006). We will then describe the Traitspace model, which incorporates trait variation within species by applying Bayes' theorem to predict species relative abundances (Laughlin et al. 2012). We will provide mathematical descriptions of the models coupled with conceptual illustrations. All readers should be able to come away with an intuitive appreciation of the models, and some may also be ready to apply them to new data sets. All the R code and data needed to apply these models and reproduce the examples are included in the GitHub repository (https://github.com/danielLaughlin/CommunityEcology).

Both of these models are generalizable to any ecosystem and can accommodate any number of species, traits and environmental conditions, provided that the number of traits is less than the number of species. Why should there be fewer traits than species? Well, if we need to measure more things (traits) than we are predicting (species), our situation has probably not improved. Also, including more traits than species almost guarantees that there is no feasible mathematical solution, and therefore no prediction at all.

COMMUNITY ASSEMBLY BY TRAIT SELECTION

Bill Shipley was the first ecologist to find a solution to the central challenge posed in this book. Shipley's prowess in mathematical modelling and plant physiology made him the perfect scientist to tackle this challenge, but the solution did not present itself immediately. In fact, Shipley had puzzled over this problem since his days as a doctoral student at the University of Ottawa. From the beginning, Shipley suspected that models of statistical mechanics developed by physicists could be useful. Shipley knew that probabilities of states could be predicted from the aggregate properties of those states and wondered if these could be applied to species. His breakthrough came in 2006, when he and his colleagues published the CATS model, which takes two basic inputs, a matrix of species trait values and a matrix of community-weighted mean trait values, to produce a vector of species abundances (Shipley et al. 2006, Shipley 2010).

Our goal here is to explain how we get to the prediction of the community vector: a vector of species abundances. To do this, we need to calculate a vector of *probabilities*, where each species in the species pool is assigned a probability. These probabilities sum to 1 within the vector across the set of species, so naturally these probabilities can be interpreted as predicted relative abundances of each species in the community vector **C** (recall Figure 1.1).

First, we need to explain the necessary input data. This input includes a matrix of species trait values that contains average trait values for each species in the species pool. This means that the user must also predefine which species are included in the pool. Recall from Chapter 3, this is not a straightforward task. The size of the species pool has important effects on the performance of the model (Sonnier et al. 2010, Merow et al. 2011). In general, increasing the size of the species pool reduces the performance of the model, but this depends also on the number of traits used as predictors. Part of the reason for this complication lies in the requirement that the probability vector sums to 1 over a set of species. Therefore, with few species, the model can assign larger probabilities to each, but inevitably with larger numbers of species in the pool each probability value must be smaller. In other words, the probabilities are diluted when the species pool is large. This is an interesting technical problem that has no simple solution at present.

How should the species pool be defined when applying these models? Recall from Chapter 3 that there are several types of pools. Both models were developed to predict how species are sorted along environmental gradients based on the matching of their traits to the environment, so the pool will be unfiltered. Ideally, the species pool should include as many species as possible that could possibly occur in your sample, that is, the regional species pool. In practice, however, we generally include (1) all species that occur in the sample of quadrats or relevés that are available across the environmental gradient, and (2) the species for which trait information is also available. The latter is most often what limits the analysis.

The models have no upper limit for the number of species that can be included. Pool size has ranged from between 15 to 506 species (Sonnier et al. 2010, Laughlin et al. 2011). Needless to say, the model performed much better on the low end of this range. If the number of species is large, and these species can be binned into a smaller number of functionally similar groups, then the model could be applied to the functional groups instead to simplify the problem. We will say more on this in Chapter 8. Now to the details of the CATS model.

The CATS model predicts a vector of species relative abundances by solving a system of linear equations and selecting the solution that maximizes the evenness of the distribution of species relative abundances (Shipley et al. 2006). We will explain both of these components in turn. The first linear equation is a purely mathematical constraint. There are no traits in this equation. This equation simply constrains the abundances of species to sum to 1. That is,

$$\sum_{i=1}^{S} p_i = 1, \tag{7.1}$$

where p_i are the unknown species, relative abundances for which we want to solve. You can see an example in Figure 7.1.

The second linear equation incorporates the traits, and takes the general form

$$\sum_{i=1}^{S} t_{ik} p_i = \overline{T}_k, \tag{7.2}$$

where t_{ik} is a vector of trait means for each species i, S is the number of species in the regional pool and $\overline{T}_k$ is the average trait value in the community (for trait k). You can see an example in Figure 7.1. There are K of these equations. In other words, we set up Equation 7.2 for each trait in the analysis. For example, if we include two traits in the model ($K = 2$), say both aerenchyma and height, then there is one Equation 7.2 set up for aerenchyma and another Equation 7.2 set up for height. Note that because this is a system of linear equations, traits must be either continuous or binary; the model cannot accommodate categorical traits with more than two levels, such as life form or functional group.

The algebraic model can be illustrated geometrically (Figure 7.1). Consider the problem of using leaf nitrogen concentration to predict the relative abundances of three species. This is a greatly simplified example for the purpose of illustration. Let the vector of leaf nitrogen concentrations (%) of each species be $\mathbf{T} = [0.5, 1.5$ and $2.5]$. What are the predicted abundances of these species, let us call it the vector $\hat{C}$, if the optimal leaf nitrogen concentration of the community, given the environment, is 2 percent? First, we must constrain the abundances to sum to 1 using Equation 7.1. That is, $p_1 + p_2 + p_3 = 1$.

This reduces the set of possible solutions from all points in a three-dimensional space called $\mathbb{R}^3$ down to a triangular plane. This plane is illustrated by the grey triangular plane in Figure 7.1, constrained by the vertices $\{1,0,0\}$, $\{0,1,0\}$ and $\{0,0,1\}$. Second, we must further constrain the linear combination of unknown species abundances (p_i) and nitrogen concentrations of each species to equal 2 percent nitrogen. Using Equation 7.2, $0.5p_1 + 1.5p_2 + 2.5p_3 = 2$.

Note that we have specified that $\overline{T} = 2$. We must either know $\overline{T}$ a priori, or predict $\overline{T}$ using environmental information in order to use this model (more on this below). By specifying this second trait-based equation, we have reduced the solution set down from the triangular plane to a single line (see the black line in Figure 7.1). However, there are still many possible solutions to this system (an infinite number actually), since the set of possible solutions includes every point along this single line!

Which specific solution, from this set of potential solutions, should we choose as our prediction for the community vector? The CATS model uses the solution that maximizes a third equation, the entropy function, where,

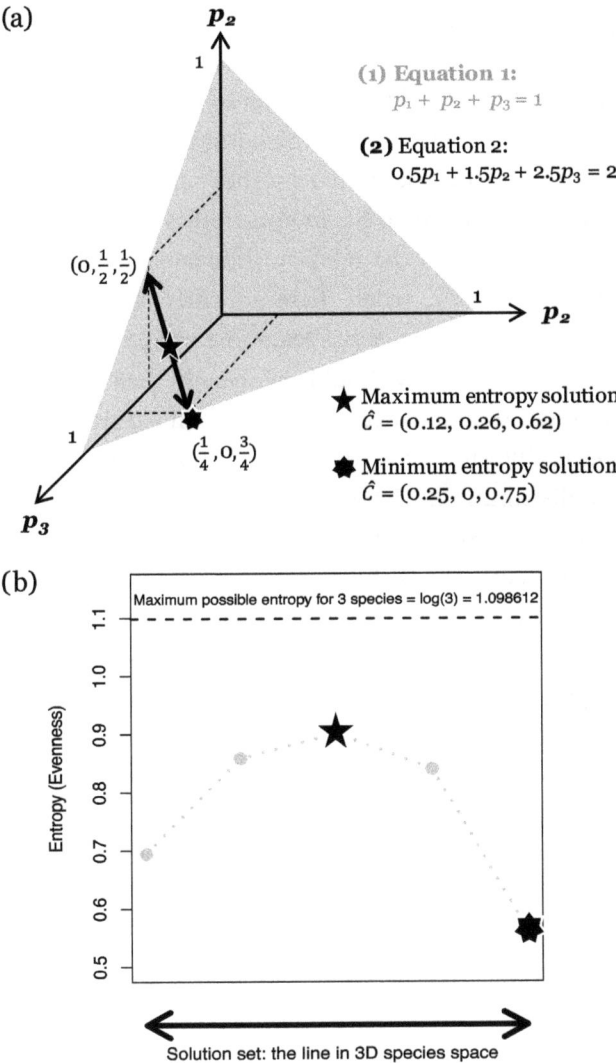

(a)

p_2

1

(1) Equation 1:
$p_1 + p_2 + p_3 = 1$

(2) Equation 2:
$0.5p_1 + 1.5p_2 + 2.5p_3 = 2$

$(0, \frac{1}{2}, \frac{1}{2})$

1   p_2

★ Maximum entropy solution
$\hat{C} = (0.12, 0.26, 0.62)$

✸ Minimum entropy solution
$\hat{C} = (0.25, 0, 0.75)$

1

$(\frac{1}{4}, 0, \frac{3}{4})$

p_3

(b)

Maximum possible entropy for 3 species = log(3) = 1.098612

Entropy (Evenness)

Solution set: the line in 3D species space

FIGURE 7.1 (a) Geometric representation of CATS for predicting the relative abundance of three species using just one trait, mean leaf nitrogen concentration per species. The first equation constrains the solutions to sum to 1, illustrated by the triangular grey plane in $\mathbb{R}^3$. The second equation constrains community-level leaf nitrogen concentration to be 2 percent, which, combined with equation 1, creates a solution set forming the black line with arrowheads. The chosen solution is the one point along the line that has maximum species evenness (★). The minimum entropy solution is (✸). (b) The entropy (evenness) of the relative abundance distribution is convex and reaches its maximum in the middle of the solution set (★).

$$H' = -\sum\nolimits_{i=1}^{s} p_i \log p_i. \tag{7.3}$$

This function maximizes the evenness of the relative abundance distribution of species. Here, log is the natural logarithm. H' has been used by ecologists for decades under another name: the Shannon diversity index. When there are no linear constraints to contend with, this index is maximized when the relative abundance distribution is perfectly even. For example, when there are three species, then H' is maximized when each species has the same relative abundance, that is, $\hat{C} = [1/3, 1/3, 1/3]$. Why should evenness be maximized? After all, virtually all relative abundance distributions in nature are dominated by a few species (Preston 1962, Whittaker 1965), which is the opposite of an even distribution. Using the logic of statistical mechanics of physical systems, Shipley argued that the maximum entropy solution is justified because it is the distribution of p_i that can be realized the most ways; therefore, it is the most statistically probable distribution. It is also the only distribution of p_i that is determined entirely from the specified constraints, without implying additional constraints (Jaynes 2003, Shipley 2010, Laughlin and Laughlin 2013).

This decision to maximize evenness is troubling to most empirical ecologists. However, it is important to realize that CATS model predictions are not actually even relative abundance distributions, despite the fact that evenness is maximized. This may seem paradoxical. It happens because the linear constraints are doing most of the work, limiting the set of possible community vectors to optimize. To explore this a bit further, let us evaluate the entropy function at multiple locations along the solution vector and compare solutions where one maximizes evenness and the other minimizes evenness (Figure 7.1b).

First, the solution that *maximizes* the entropy function in our example of three species is the vector $\hat{C} = [0.12, 0.26, 0.62]$ (Figure 7.1). You can reproduce this answer yourself using the R code provided in the GitHub repository. Note that this is not an even distribution of relative abundances, despite the fact that we maximized evenness! The third species is predicted to be dominant.

There are two other things to notice in Figure 7.1b. First, the maximum possible entropy for any given number of species, S, is log (S). No solution ever reaches the maximum possible entropy value of $log(3) = 1.098$. That is because the linear constraints must be satisfied. The solution {0.33, 0.33, 0.33} is not a solution to this set of linear equations. Second, we also notice that entropy is convex such that entropy is maximized toward the middle of the solution set (Figure 7.1b), and declines to either side. From a technical perspective, this means that most optimization algorithms can get trapped in local minima if they are seeking the solution that minimizes entropy. (For keen readers, we demonstrate this problem in the R code.) To find the global minimum, one needs to evaluate the function at multiple random locations to ensure one is finding the true minimum.

Second, the solution that *minimizes* the entropy function in our example is the vector $\hat{C} = [0.25, 0.00, 0.75]$ (Figure 7.1). It is possible to get trapped in the local minimum at the vector $\hat{C} = [0.00, 0.5, 0.5]$, but this relative abundance distribution actually has a higher entropy value than the true minimum (Figure 7.1). These vectors clearly illustrate that the trait-based constraint (Equation 7.2) exerts far greater control over the relative abundance distributions than the entropy function itself (Equation 7.3). Therefore, whether we maximize or minimize entropy, the resulting prediction is a community dominated by just a few species, similar to what we see in nature.

This leads to a second issue. We have explained how it is that the model chooses a community dominated by one or two species. But how does it determine which of the species will be the dominant? In our example, the third species is predicted to be dominant in both cases. Why is this?

Here is the key point: the system of linear equations assigns the largest relative abundances to the species with trait values that are closest to the community-weighted mean (CWM) trait constraint. In other words, species that have trait values that are closest to the optimal trait value (i.e., the CWM trait value) should dominate the community. This is common sense: the community consists of species with the best fit to tolerate the filters acting at that location.

Note that this model is purely a translation of environmental filtering and ignores the complexities of population dynamics and competitive interactions. According to Shipley et al. (2006), their

> model ignores, but does not deny, the details of resource allocation, population dynamics, stochastic processes, and species interactions. It assumes a constrained random allocation of resource units to species; the constraints are generated by natural selection and are quantified by the community-aggregated trait values. The relative abundance of each species in a species pool is therefore a function of how closely its functional traits agree with the community-[weighted mean] traits.

This formulation means that if several species share similar trait values that are all close to the optimum, then all these species will dominate the predicted community vector. This contrasts with the theory of limiting similarity where species that differ from each other are the species that can coexist in a community. How local-scale competition integrates with trait-based environmental filtering is a topic we will revisit in Chapter 8.

The CATS model has been shown to predict significant variation in species abundances based on their functional traits in a growing list of ecosystems, including old fields and upland rangelands in Europe, montane and subalpine forests in the southwestern USA, Ponderosa Pine forest understories in Arizona, arid steppes in North Africa, tropical forests in French Guiana and tussock grasslands of New Zealand (Shipley et al. 2006, 2011, 2012, Sonnier et al. 2010, 2012, Laughlin et al. 2011, Merow et al. 2011, Laliberté et al. 2012, Frenette-Dussault et al. 2013). As far as we are aware, all studies to date have been conducted in plant communities, but we hope these tests are also conducted in invertebrate, fungal, fish, bird, mammal and microbial communities as more trait data become available.

Let us look at one example. We discussed drought as a strong environmental filter in Chapter 2. Let us turn to the arid steppes of North Africa to see how aridity drives community assembly (Frenette-Dussault et al. 2013). These steppes are near the Atlas Mountains of Morocco, further south than the Baetic–Rifan biodiversity hotspot discussed in Chapter 5 (Figure 5.5). There are cold winters and hot dry summers, with <250 mm of rain annually, and the landscape, already arid, is being degraded from grassland to spiny shrubs as a result of overgrazing of sheep and goats. For context, Thirgood (1981) describes the history and impacts of overgrazing in the Mediterranean region overall, while Wright (2017) explores potential grazing impacts in North Africa as a whole. The data consisted of 50 plant communities, with 14 functional traits measured on the most dominant 34 species. The CATS model was applied to test predictions of relative abundances, and over 90 percent of the variation in observed species relative abundances were explained by the model when observed CWM traits were used as constraints (i.e., $\overline{T}_k$) in Equation 7.2.

The use of observed CWM traits as $\overline{T}_k$ in the model is open to criticism because community structure encoded in $\overline{T}$ has to be measured before species abundances can be predicted, making the model appear circular (Marks and Muller-Landau 2007, Roxburgh and Mokany 2007). However, both common sense and ecological theory suggest that average trait values in a community are likely under selection by the environment (Enquist et al. 2015). We explore many examples of trait–environment relationships in Chapters 2, 4 and 5. Therefore, it is reasonable to conclude that $\overline{T}_k$ can be predicted using environmental factors. Getting back to North Africa, many of the community-level traits in the Moroccan steppe were significantly correlated with aridity. The proportion of succulents, the onset of flowering and the proportion of woody species all increased with aridity. In other words, aridity filtered the species pool by selecting species with traits that permitted their survival under extreme water limitation. This means that the CWM traits could be predicted using environmental information rather than measuring the CWM traits directly. Regression models would be one obvious way to predict CWM traits from environmental variables, as we explored in Chapter 5.

When predicted CWM traits, as opposed to observed CWM traits, were used in the CATS model, 40 percent of the variation in observed abundances was still explained by the model (Frenette-Dussault et al. 2013). Using predicted rather than measured CWM traits eliminates any apparent circularity and several other studies using this approach made significant predictions of species abundances (Merow et al. 2011, Shipley et al. 2011). It is also possible to argue that using the observed traits is quite acceptable, on the grounds that regression models also fit models to data that have been measured before fitting the model (Shipley 2010). In fact, CATS has been shown to be equivalent to Poisson regression (Warton et al. 2015). If CATS is considered to be just another form of regression, then using observed values may be justified to determine which traits are important in community assembly.

The prediction of CWM traits from environmental filters is useful not just for testing CATS predictions, but also for making wider predictions, such as expected combinations of traits and functional types in new sites or in new environments. This is the ultimate test of the model. The model can be applied to forecast responses of ecological communities to changing climates. Two climatic factors, temperature and aridity, are likely of broad general interest, and have already been documented as important environmental filters in both the arid steppes of Morocco (Frenette-Dussault et al. 2013) and the montane and subalpine forests of the southwestern USA (Laughlin et al. 2011). If climate can predict the average trait value in a community (which seems probable), and if projections of future climate are available, then future projections of climate can be used to predict changes in traits and, therefore, communities. These predicted community-level traits can then be used to project changes in species abundances over time in response to climate change. And, as we will observe in Chapter 8, such information can guide habitat restoration.

HOW MANY TRAITS ARE ENOUGH?

We have already posed the question of just how much data we need in community ecology, a question that Tansley posed more than a century ago (Tansley 1914), and an issue that we raised in Chapter 5. This issue arose in a new guise in Chapter 6, when we asked how many functional groups are needed to describe a community. With measurement, the answer seems to be that it is always easy to measure more trivial things, and this tendency needs to be discouraged. With functional groups, the answer may be similar. We need to avoid unnecessary effort and complexity. We need to keep the number rather small, accepting the limits of human minds. Of course, computers can make models using ever more functional groups, as many as one wants, really, if the computer is big enough, but if humans cannot comprehend the results, what is the point? Some people are easily impressed by big models, as if a big complex model has some inherent value. This tendency can distort how we think about solving problems in ecology (Keddy 2005). Bigger is not necessarily better, even in the world of pickup trucks. In most cases, fewer than 10 functional groups is usually a good start. In this context, let us ask how many traits are needed? With traits, it likely has something to do with the degree of correlation among traits: when is a new trait bringing new information, and when is it so strongly correlated with other traits that the effort is unnecessary? This in turn is related to how many fundamental axes of variation there are in trait space. Figure 4.7 suggested there are at least seven possible axes in trait space for plants. If we add more traits, there is not only a waste of effort, but a risk of "overfitting" (Christian and Griffiths 2016).

In this context, let us ask how many traits we need to put into a CATS model for satisfactory results. It turns out that one can answer this question with a good deal of precision using meta-analysis of studies that tested the CATS model (Laughlin 2014). In every model, the predictive power increased rapidly with the addition of traits, but this predictive power levelled off asymptotically (Figure 7.2). This general result has two implications. First, multiple traits are important in community assembly, and predictive power is optimized by including traits from different plant organs (leaves, stems, roots, flowers). For example, leaf, root and seed traits were most important in the Ponderosa Pine forest understory of northern Arizona (Shipley et al. 2011). In the fynbos of South Africa, leaf traits, stem traits and flowering phenology were most important (Merow et al. 2011). Combinations of traits that reflect different adaptations and provide unique information about the functional differences of species are important for maximizing our understanding of trait-based environmental filtering. This is the general lesson of Figure 7.2, and it likely applies to all trait matrices. If you are studying bird community assembly, it is best to include traits on bills, wing dimensions, talons and other organs. When studying fish community assembly, it is best to include fin and body morphology, thermal maxima and properties from other organs. This is one reason why there is often merit in physiological attributes such as nitrogen content of leaves or cavitation resistance in xylem.

The second implication of Figure 7.2 is that there is an upper limit to the number of traits that are useful for making predictions. The curves in Figure 7.2 suggest that the ability to predict community composition stops increasing substantially after 4–8 traits. The reason for this is that many traits are correlated, and including multiple correlated predictors in any model does not yield dividends. Including multiple correlated traits does not add unique information about each species, suggesting that including more than eight traits leads to diminishing returns. When applying these trait-based models, the lesson is clear: ecologists need to minimize the number of traits while maximizing the number of unique functional dimensions (Laughlin 2014).

THE TRAITSPACE MODEL OF ENVIRONMENTAL FILTERING

The CATS model has proven to be a relatively direct approach for predicting a vector of species abundances from a species pool using

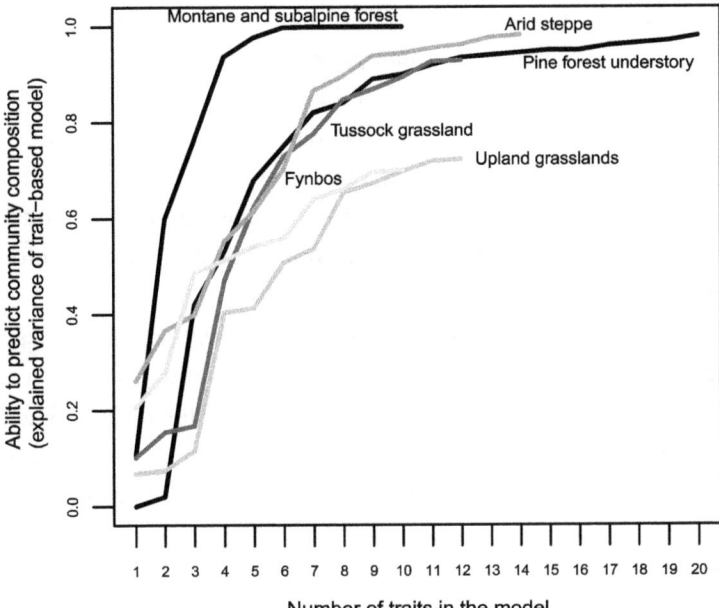

FIGURE 7.2 The relationship between the number of traits and the ability
to predict community composition (based on the R^2 of the relationship
between observed and predicted relative abundances) using the CATS
model of community assembly (adapted from Laughlin 2014, using data
from Laughlin et al. 2011, Shipley et al. 2011, Merow et al. 2011, Sonnier
et al. 2012, Laliberté et al. 2012, Frenette-Dussault et al. 2013).

functional traits. Still, there are some limitations. First, it is
limited to species average trait values, but even a single species
can show great variation in its functional traits. We have not had
a good deal to say in this book about phenotypic variation within
a species, mostly because it is typically small relative to interspecific phenotypic variation. Second, the CATS model depends on
CWM traits, and the use of a single CWM trait to describe
a community assumes that there is a single optimal trait value
for the community. Yet, we know that multiple phenotypic solutions have evolved to solve the same environmental filter, even
strong filters like drought (recall Figure 2.1) and flooding (recall

Figure 6.13). Continuing with drought, in arid ecosystems like the one shown in Figure 5.5, fast-growing annual plants avoid drought by growing in the brief cool and rainy season, whereas sclerophyllous shrubs tolerate drought by maintaining photosynthesis when water potentials are low, often by growing deep roots that tap water far into the ground. These strategies coexist. Daniel and his colleagues developed the Traitspace model as a means to address both of these limitations.

Just like CATS, Traitspace predicts a vector of species relative abundances based on the matching of their traits to a given environment (Laughlin et al. 2012, 2015, Laughlin and Joshi 2015). The name of this model refers to its reliance on multidimensional distributions of species within a mathematical "space" defined by traits. If one trait is used for prediction, then this space is one-dimensional. If two traits are used, then it is two-dimensional, and so on. Think of a trait matrix: while CATS uses a matrix of trait values where each row is a different species, Traitspace uses a matrix of trait values where each row is a different *individual*. This allows us to include the variation of each trait both within and across species. Further, while CATS does not strictly rely on a matrix of environmental variables, Traitspace requires that for every trait observation there must be a corresponding environmental variable. Finally, Traitspace does not require maximizing the entropy function, so it makes no *a priori* assumption about which distribution is most likely. Now for the model itself.

Traitspace is grounded in a causal graph model that states that the relationship between the community vector of species abundances (C) and the environment (E) is mediated through functional traits (T). In other words, $E \rightarrow T \rightarrow C$ (Figure 7.3). Traitspace predicts the relative abundance of every species from the species pool in a given environment. We denote this symbolically as $P(C_i|E)$, which can be read as the "probability of each species given the environment." On the surface this looks like the output of standard species distribution models where species ranges are only a function of the environment, not their traits. However, in the Traitspace model the predicted relationship between species and environment is determined by whether or not the traits are a good match to the environment. The traits are the engine that does most of the work.

Conceptually, Traitspace uses information encoded in the trait–environment relationship as a measure of the environmental filter (recall Proposition 4). Consider the illustrated hypothetical example in Figure 7.3 (upper right), where the relationship between traits and the environment is strong. Then, at a given point along the environmental gradient (a specific habitat if you like, denoted as the region between the arrows in Figure 7.3), only a certain range of trait values will be likely to occur there. Species whose traits align with this range will be given the highest probability of occurring in that habitat. The species that are denoted by the darkest shades of grey exhibit the greatest overlap with the environmental filter, so these species are

Conceptual diagram of *Traitspace*

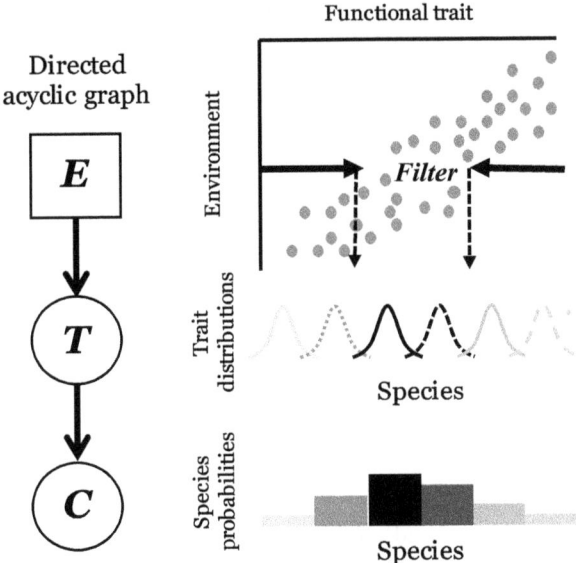

FIGURE 7.3 Conceptual diagram of the Traitspace model (adapted from Laughlin et al. 2012).

given the highest probability of occurring in that habitat, producing the histogram at the bottom of Figure 7.3.

The Traitspace model has two stages: a *calibration* stage and an *inference* stage. We illustrate the Traitspace model framework using a simple simulated example (Figure 7.4), which you can reproduce for yourself using the R code in the GitHub repository. The calibration stage itself has two steps. Each step generates information that we will eventually combine using Bayes' theorem to predict species abundances. First, we characterize the environmental filter using a generalized linear model to fit $T = f(E)$. It may be easiest to think of this step with simple linear models, but it is equally possible to use more complex nonlinear models, depending on the complexity of the relationship. The fundamental point is to generate prediction intervals for individual-level (not community-level) trait values based on environmental variables.

To begin, Figure 7.4a illustrates a positive linear relationship between one trait and one environmental gradient. Examples might include leaf area versus rainfall, bark thickness versus fire frequency or spine length versus grazing pressure. This calibrates the conditional distributions of individual-level traits. By "conditional" we mean the trait distributions that are likely given the environmental conditions, which we denote symbolically as $\varphi_{T|E}$. In other words, it is more likely to observe a high trait value at the high end of the environmental gradient. This approach differs from CATS because it models variation in individual-level traits, as opposed to variation in CWM traits. Because of this, the prediction intervals of these regression models can be quite large, given the wide range of trait values that are typically found within a diverse community. So the first step yields a regression model that quantifies how the mean and the variance of a trait changes across an environmental gradient. If there is no relationship between a trait and the environmental filter, then the trait will be useless for predicting community assembly. If, however, there is a predictable change in traits along the environmental gradient,

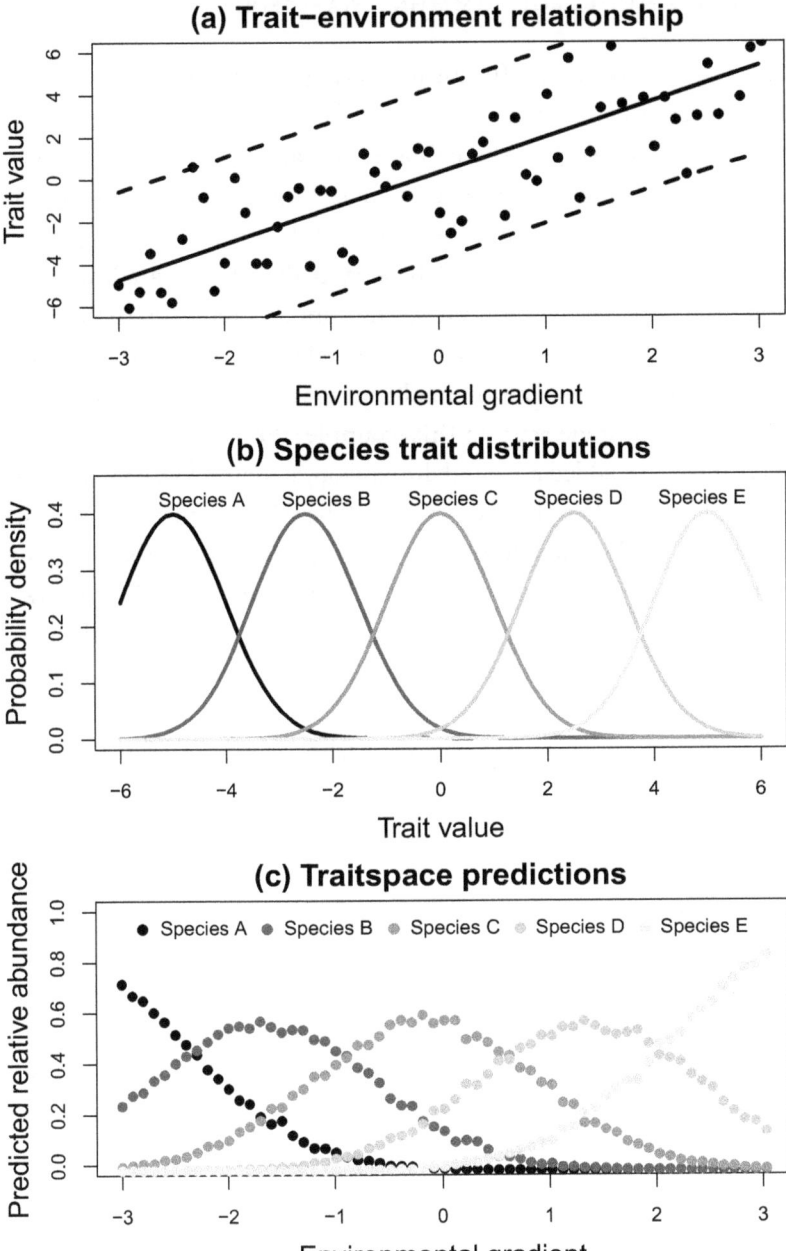

FIGURE 7.4 Illustration of the Traitspace model for predicting the relative abundance of five species: (a) simulated positive relationship between traits and environment; (b) probability distributions of the trait for each of the five species; and (c) Traitspace predictions using the information in panels (a) and (b). Species A has a high probability at the low end of the environmental gradient because it has low values of the trait. The opposite is true for Species E.

then the trait may be useful for predicting changes in species abundances across the gradient.

The second step of the calibration stage is to characterize the location and dispersion of species in n-dimensional trait space to calibrate the conditional distributions of traits for each given species, denoted symbolically as $\varphi_{T|C_i}$. These trait distributions allow us to differentiate species by their traits. The simplest approach to this step is to assume that traits are normally distributed within each species. Figure 7.4b illustrates hypothetical trait distributions for each of five species across a single trait. In this example, the trait space is a single dimension. If we added a second dimension, which we will do in a real example later on, then the trait space will be two-dimensional. We call these trait distributions "probability density functions" because the area under each curve sums to 1. What do these trait distributions tell us? Figure 7.4b demonstrates that if you are considering the light-grey species E, the probability of observing a low trait value is near zero, whereas the probability of observing a large positive trait value is high. As a practical example, the figure might show the probability of having sclerophyllous leaves or aerenchymous roots. So after completing the second step, we can locate each species within a mathematical region called trait space. If species overlap in this trait space, they are functionally similar and may be adapted to similar environments. If they are located in different regions of trait space, then they are functionally different, and may be adapted to different habitats.

We are now armed with two key sources of information to predict the community vector $\mathbf{C}$. But before we do so, let us take a little more time to explore what this process means, and come to our own reasoned conclusion without using mathematics. We know from Figure 7.4a that it is more likely to observe a high trait value at the high end of the environmental gradient. We also know from Figure 7.4b that we are likely to observe a high trait value if we are observing the light-grey species E. Taken together, logic tells us that we are more likely to observe the light-grey species E near the high end of the environmental gradient. Now that we have developed an intuition about our prediction, let us do the calculations.

Following the calibration stage, we begin the inference stage. Inference involves the following four steps. First, we simulate community assembly stochastically by generating a large number (N) of simulated trait values using the mathematical distributions $\varphi_{T|E}$ at every position along the environmental gradient. In other words, the output of this step of the model is a set of trait samples that can possibly occur within a given habitat, and these trait samples are the raw material for the following steps. Next, for every simulated trait value, we calculate the likelihood $P(T \mid C_i)$, that is, the probability of a trait value given the species, using the conditional distributions $\varphi_{T|C_i}$. Third, for every trait value, we compute the posterior distribution of species conditioned on both the trait data and the environmental conditions using Bayes' theorem (McCarthy 2007). This powerful theorem is named after the Reverend Thomas Bayes (1701–1761), who was the first to use conditional probabilities to calculate limits on an unknown parameter. It has been said that this theorem "is to the theory of probability what the Pythagorean theorem is to geometry" (Jeffreys 1973). The use of this theorem in ecology has risen exponentially with the rise of computational power, and we use this theorem to compute the posterior distributions.

The posterior distribution is denoted as $P(C_i \mid T, E)$, which can be read as the "probability of each species given the trait value and the environmental conditions." $P(C_i \mid T, E)$ is computed using Bayes' theorem:

$$P(C_i|T,E) = \frac{P(T|C_i)P(C_i)}{\sum_{i=1}^{S} P(T|C_i)P(C_i)}. \tag{7.4}$$

$P(C_i)$ denotes a flat or uniform prior on the species. The flat or uniform prior can be considered a kind of null model, the highly unrealistic assumption that all species are equally likely to occur in this habitat, a situation that arises when we have no prior knowledge. If we suspect *a priori* that certain species are more likely to occur (e.g., some species were more abundant in a particular habitat historically whereas others were not), then more informative priors could be entertained (Shipley et al. 2012). For simplicity and objectivity, we will stick with the uninformative prior. The numerator in Equation 7.4 is the product of the likelihood and the prior, and the denominator is a normalizing term so that the probabilities across all species sum to 1. Note that Equation 7.4 is valid because, according to the implications of the causal graph model (Figure 7.3), we have $P(C_i|T,E) = P(C_i|T)$.

We are almost finished. Remember our original goal was to compute $P(C_i| E)$, the probability of each species given the environment. This means that we have to use integration to extract the traits from $P(C_i| T, E)$. The desired information is computed by integrating with respect to traits to obtain the relative abundances of species given the environmental conditions:

$$P(C_i|E) = \int P(C_i|T, E)P(T| E)dT. \tag{7.5}$$

This tricky integral can be adequately approximated using Monte Carlo integration as

$$P(C_i|E) \cong \frac{1}{N}\sum_{k=1}^{N}P(C_i|T_k, E)P(T_k| E). \tag{7.6}$$

Monte Carlo integration effectively computes the predicted vector of abundances, C (Figure 1.1), in a given habitat by averaging across all the simulated vectors generated during the inference stage. That is to say, at the beginning of this process we have a cloud of community vectors, but at the end we have achieved our goal of predicting a single vector, giving us the most likely community vector in this habitat.

Recall that in Figure 7.4 we used our intuition to predict that we would be more likely to observe the light-grey species E at the high end of the environmental gradient. We have now confirmed our intuitive prediction using the logic of probability! The positive trait–environment relationship suggests that species with high trait values will be more likely to occur at the high end of the environmental gradient (Figure 7.4a), and the light-grey species E exhibits the highest trait values (Figure 7.4b). Consequently, the light-grey species E has the highest probability of occurring at the high end of the environmental gradient (Figure 7.4c). Each of these gradients can be considered to have many habitats, each determined by the causal factors acting at that point along the gradient. In this sense, the environmental gradient is like a row of different habitats.

Now that we have illustrated the machinery of Traitspace using idealized simulated data, let us see how the model performs when applied to the real world.

APPLYING THE TRAITSPACE MODEL TO CONIFER FORESTS IN SOUTHWESTERN NORTH AMERICA

The first test of this model was conducted along a large climatic gradient in the southwestern USA, using data on tree species composition from 196 plots in the mountains near Flagstaff, Arizona (Laughlin et al. 2012). These forests range from lower elevation semi-arid Ponderosa Pine forests where low-intensity fire is frequent, up through cooler mixed conifer forests where fire is more sporadic in time and intensity, all the way up to the timber line where long-lived Bristlecone Pine (*Pinus aristata*) trees survive the harsh cold temperatures in the absence of fire. In addition to the species composition data in the 196 plots, three traits were measured on trees from across the climatic gradient. Thus, the required input data for running the Traitspace model were: (1) trait values measured on *individual* trees at sites with known climatic conditions, and (2) environmental conditions (e.g., mean annual temperature) at each site where species composition was observed in order to test the model predictions.

Before moving on to the results, we note that two of these traits proved to be particularly powerful for predicting species abundances: wood density and bark thickness. In general, wood density is related to the ability of a tree to tolerate drought conditions (Hacke et al. 2001). In this particular study, wood density strongly increased nonlinearly with mean annual temperature. High wood density was found in the most arid end of the gradient, but also in the alpine zone where plants experience physiological drought because water is frozen much of the time. You will recall from Chapter 2 that drought resistance is a key factor in tree distributions, and that the mechanism appears to be the embolism resistance of the xylem (Maherali et al. 2004, Choat et al. 2012, Larter et al. 2015). You will also recall from Chapter 2 that wildfire is a strong filter on forest ecosystems globally. In general, bark thickness is related to a species' ability to tolerate low-intensity surface fires because thicker outer bark (i.e., the visible layer that is produced outward by the cork cambium) provides an insulating layer to prevent heat-induced damage to the vascular cambium (Pausas 2015). In the Arizona mountains, bark thickness increased exponentially with increasing temperature. Warmer, drier forests experience a higher frequency of low-intensity surface fires, and for trees to persist in such an environment, they must insulate their vascular cambium from heat damage.

The Traitspace model produced predictions of abundances for the nine dominant tree species, and these predictions closely resembled observations in the field (Figure 7.5). Moreover, the predicted optimal temperatures for each species were similar to (and closely correlated with) the observed temperature optimums (Laughlin et al. 2012). Consider five tree species. Bristlecone Pine has dense wood and thin bark, and was therefore predicted to occur at the highest elevations, where freezing-induced cavitation is common and where fires are infrequent. Engelmann Spruce (*Picea engelmannii*) and Subalpine Fir (*Abies lasiocarpa*) have thin bark and low wood density, and these species occurred at intermediate elevations, where drought and fire were less frequent. Douglas Fir (*Pseudotsuga menziesii*) and Ponderosa Pine (*Pinus ponderosa*) have moderately dense wood and the thickest bark, and were predicted at the low end of the environmental gradient, where both drought and fire were frequent.

Readers might, at this point, be feeling overwhelmed, particularly those without previous familiarity with linear equations or Bayesian models. So this is a good time to acknowledge that we have

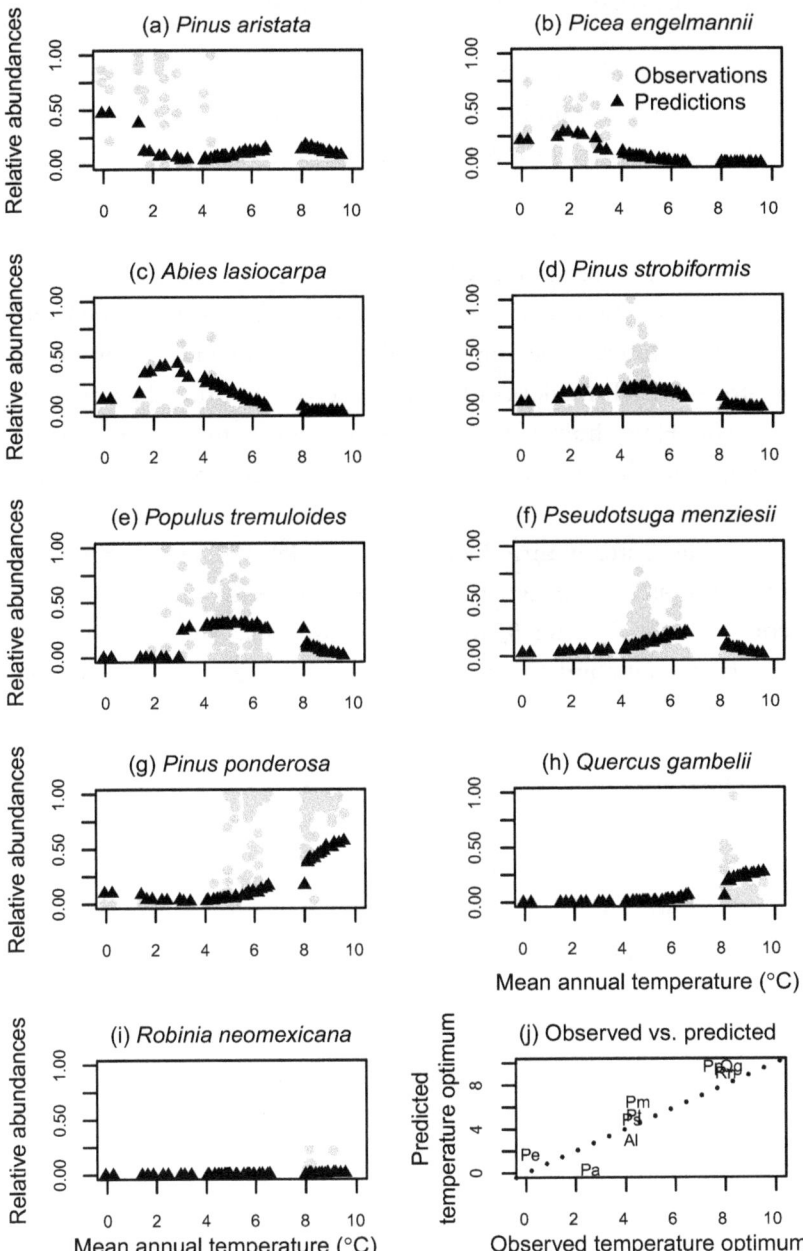

FIGURE 7.5 (a–i) Comparison of observed species abundances (grey circles) to those predicted by the Traitspace model (black triangles) for nine tree species along a temperature gradient in the southwestern USA. (j) Comparison of observed and predicted temperature optimums for each species. Initials represent the first letter of each genus and species; the dotted line is the 1:1 line (adapted from Laughlin et al. 2012).

already covered a lot of ground in this chapter, introducing two different models for community assembly. It is entirely understandable if you feel the need to put the book down and take a break. Before you do, allow us just a brief summary. We have introduced two models that assemble a community vector (**C**) from the species pool (**P**). Both therefore arise quite naturally from the very first figure in this book, Figure 1.1. Both models use data on traits to accomplish this task, but make rather different assumptions about the constraints that govern this process. Each model could even receive an entire chapter in its own right. Here, however, we have decided to introduce them in tandem, to emphasize their common starting point and their common purpose.

The next step in introducing these models will be to apply them both to the same data set, in this case the kettlehole wetland you encountered in Chapter 5, shown in Figure 5.3. All the same, this might be a good point to put on the kettle, break for tea and let the mathematics settle into place.

APPLYING BOTH CATS AND TRAITSPACE TO A KETTLEHOLE WETLAND

Let us return to the kettlehole wetland in New Zealand (Figure 5.3) and apply both models to the same ecosystem that is structured by a strong gradient in soil hydrology. This will show us how both CATS and Traitspace approach the challenge of predicting the outcome of community assembly. You will recall that root aerenchyma and height varied significantly along the flooding gradient, but the analyses described in Chapter 5 determined that only aerenchyma was driving the environmental filtering.

How does the CATS model perform using root aerenchyma and height in the ephemeral wetland of New Zealand? As a reminder, you can reproduce these results using the R scripts located in our GitHub repository. The CATS model yielded significant predictions of species relative abundances across the flooding gradient when root aerenchyma was used in the model ($R^2 = 0.29$, $p = 0.017$), but not when height was used ($R^2 = 0.01$, $p = 0.75$).

What about Traitspace? We can apply Traitspace to this system because traits on individual plants were measured along the flooding gradient (Purcell et al. 2019). The Traitspace model yielded significant predictions of species relative abundances across the flooding gradient when root aerenchyma was used in the model ($p = 0.001$), but not when height was used ($p = 0.998$).

These two analyses demonstrate how both models distinguish between traits that are important to community assembly versus those that are not. CATS can do this even though it relies on CWM traits as constraints in the model because the tests of significance account for the "species effect" using permutation tests. Traitspace was successful while accounting for intraspecific trait variation and without relying on CWM traits. We should note that the CATS and Traitspace models will not always agree (Laughlin and Laughlin 2013). Why should they, given their fundamental differences in underlying mathematics? However, given such a strong filter (flooding) and a clear mechanism to enhance fitness in flooded soil (root aerenchyma), both models successfully predicted the outcome of community assembly.

RELAXING THE ASSUMPTION OF A SINGLE OPTIMAL TRAIT VALUE USING TRAITSPACE

Up until this point, we have used each model with the assumption that there is a single optimal trait value for a given habitat. If we view the traits within a fitness landscape (as opposed to a real landscape), this would be illustrated as a single peak within the fitness landscape (Poelwijk et al. 2007, Laughlin 2018). But as we saw in Chapter 6, there are many cases where a single habitat seems to have multiple trait solutions – that is, where the fitness landscape seems to have multiple peaks. We illustrated this with very two diametrically opposed situations: plants in wetlands (recall Figure 6.13) and plants in deserts (recall Figure 2.3). Hence, it would appear that in some trait spaces there are multiple solutions for populations to persist in these land-scapes (using landscapes in both senses of the word). The fitness landscape in these scenarios may be rugged, with multiple crags and peaks (Laughlin 2018). Can we modify the modelling approach to account for multiple coexisting functional types?

This problem can be solved by modifying the flexible Traitspace model (Laughlin et al. 2015). Let us look at an example using two important traits, leaf nitrogen and phosphorus content, arrayed along a different kind of gradient than we have discussed previously – a soil nutrient gradient found along a soil age chronosequence in New Zealand (Mason et al. 2012). The Franz Josef chronosequence in New Zealand represents a gradient of soil ages from 60 years old to 120,000 years old, based on the continual retreat of the Franz Josef glacier. Immediately following glacial retreat, total soil phosphorus can be quite high in young soil (>600 mg kg^{-1}). However, after thousands of years of uptake, loss and leaching (the wet West Coast of New Zealand can experience >6,000 mm of rain annually!), total soil phosphorus is reduced to <200 mg kg^{-1}. Mason et al. (2012) observed that leaf nutri-ent concentration converged on very low concentrations of both leaf nitrogen and phosphorus in the older soils. In other words, there was evidence for convergence of leaf trait values upon a single optimum peak in phosphorus-poor soil (Figure 7.6). The situation was very different at the other end of the gradient, in the young phosphorus-rich soil. Leaf nutrient concentrations were higher in the younger soils, but it was not just the mean that increased, the variation increased as well. Moreover, the number of optimum trait values increased with increasing soil phosphorus, as evidenced by the three to four peaks in leaf nutrient trait space (Figure 7.6).

The Traitspace model was adapted to allow for a larger number of optimal trait values within a single habitat to improve predictions of community assembly (Laughlin et al. 2015). The main difference between previous examples is the shape of the mathematical distribu-tion $\varphi_{T|E}$. Previously, this distribution was normally distributed, centred on a single optimal value. In the new adapted model, however,

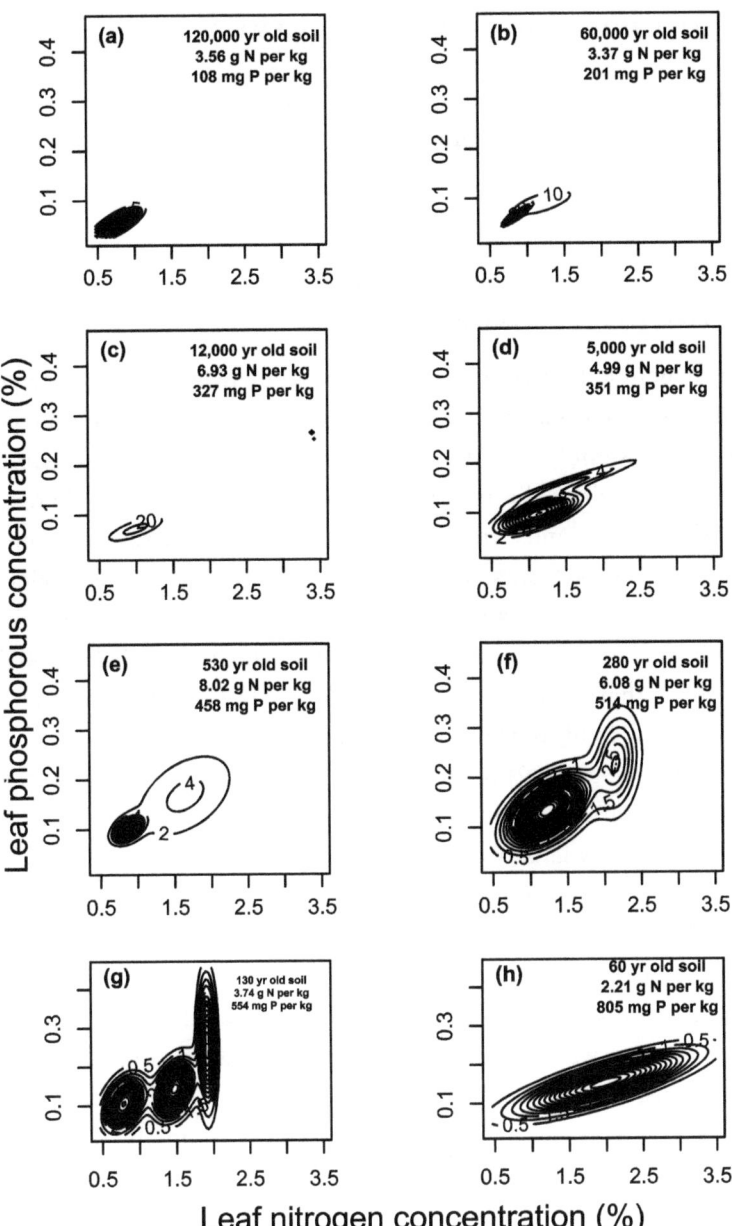

FIGURE 7.6 Bivariate contour densities of leaf nitrogen and phosphorus concentration (%) measured on individual plants within each of the eight sites on the Franz Josef chronosequence arranged from (a) oldest soil (lowest total soil P) to (h) youngest soil (highest total soil P). The older P-poor soils at the top of the figure tend to exhibit single peaks in the trait space (a–d), whereas young and more fertile soils generally exhibit broad distributions, sometimes with multiple adaptive peaks in the trait space (e–h) (adapted from Laughlin et al. 2015).

this distribution had multiple peaks and was fit to the data using mixture models. In other words, $\varphi_{T\eta E}$ can be a distribution with multiple peaks that acknowledges the fact that multiple trait optima exist (Figure 7.6).

SOME CONCLUDING REMARKS ON TWO MODELS FOR COMMUNITY ASSEMBLY

We can conclude that trait matrices (i.e., knowledge of how species vary in their functional traits) can be used to predict the outcome of community assembly in different habitats. The importance of being able to make such predictions grows by the day, as land-use change, habitat destruction and climate change threaten native species, communities and ecosystems. Both the CATS and Traitspace models make specific predictions of relative abundances for real species on real environmental gradients. Both models are pragmatic (Keddy 2005) in the sense that they can be tested using data that are commonly collected in the field, and both models seek to advance a more predictive ecology (Peters 1992) because, with appropriate input data, they can make predictions for any community on the planet.

We have now largely completed our self-assigned task. We have offered a general framework for community ecology. We have explored how species pools, filters and traits provide the raw material for the assembly of ecological communities. We have also examined the relationships between traits and environmental factors, and the ways in which trait matrices provide information on functional groups. We have concluded with two models that assemble communities from species pools. Thus, our story is mostly complete.

We still have relatively few examples of these models being applied to real communities. No doubt this will change. We will learn more about the relative value of these models, and their limitations, when they are applied to a wider array of natural communities. Most of the published work comes from plants, and a good deal of that from relatively arid environments. What about predictions of the assembly of insect, fish or bird communities? We hope the models continue to be applied and tested, and the results of such tests be published. Maybe you have ideas to improve these models, or the ideas presented in this book have inspired you to develop a different model altogether that uses data on species pools and environmental filters to predict the community vector.

If you are using this book to familiarize yourself with the basic principles of community ecology, and the two predictive models that currently assemble communities from species pools, we invite you to rest here. In this case, it would likely be helpful to mentally revisit the historic foundations of ecology, and consider how the work we have shared provides an end point of sorts. In this case, the narrative begins with early ecologists observing how different communities arise in response to different environments. You might wish to revisit Chapter 1, to reinforce your understanding of the basic problems that community ecology is trying to solve. You might then revisit Chapters 2 through 4, to remind yourself of the raw materials for community ecology: environmental filters, species pools and traits. You might also wish to make a short list of key scientists whose work has been woven into our book.

Speaking of historical personages, we have reminded you that Tansley himself warned against going into the field to collect data at random. Just because you have a large budget and a helicopter, flying around collecting soil samples or doing bird counts is not necessarily justified (although, of course, it would be fun). Here we remind you that a good scientist is also a good naturalist: there is no problem in going into the field to make random observations as a kind of hobby. A morning spent in the forest to see what birds are calling, or an afternoon spent in a wetland to see what plant species are flowering, is time well spent. The point is not to denigrate nature study, it is that such activities do not necessarily qualify as science. It is helpful to keep a practical distinction between recreation and work, although both can involve visits to wild places and pleasurable field experiences. We also encourage you to keep a field notebook, so long as you understand that this is mostly a record of random events, not an exercise in scientific hypothesis testing. (Tansley also spoke about keeping lists of species.) We are confident that over a career, the practical experience of days in the field exploring natural history will accumulate to allow you to accomplish better science overall, so long as you do not confuse the two. Rob Peters (1980) also wrote about the difference between natural history and science, although perhaps not quite in the same sense as we use the words here. We will have just a little more to say about the differences between science and natural history in the concluding chapter.

If you are a younger scientist learning these techniques for the first time, we would also encourage you to keep well aware of the various environmental problems arising in the world, and to think carefully about the potential utility of your research, particularly for wild places and wild species. We have tried to weave in a few wild places into the story (e.g., Kruger National Park in South Africa, the Amazon basin rain forests, conifer forests of the Rocky Mountains and Mediterranean landscapes); many more could have been included. We hope you will be inspired toward goals that include the protection of wild species (the species pool of the entire Earth), and therefore toward the protection of the wild places in which they live.

We appreciate there is a second audience, readers who are already familiar with some of the material in this book, scholars who are actively seeking to advance the field further. Let us turn our attention to you. We will have more to say to this audience in our concluding chapter, titled "Prospects and Possibilities."

Our first piece of advice for experienced scientists is, however, also useful for beginners: familiarize yourself with historical foundations. The past provides a vital reference point for modern work, since it clearly indicates the direction in which community ecology is moving. It allows us to see directionality in the mass production of publications and conference presentations. It also allows us to make vital course corrections if the study of ecology seems to be moving in the wrong direction. You cannot assume that moving in a random direction will get you to your desired location on a hike in the mountains, or on a road trip or in a scientific discipline. In most cases, it is necessary to know both the starting point and your desired end point. Lacking these two reference points – be honest – you are lost. At no time do we want grant support paying for lost scientists to wander randomly about the fields of academe, reporting on their favourite natural history observations, interrupted only by the occasional sabbatical, while the world has growing lists of endangered species and habitats. Nonetheless, we have both heard stories about lecturers telling impressionable students that they are wasting time reading work that is more than five years old. We hope that such stories are apocryphal, like Zeno's tortoise, but we sadly think they might be real. You should pay careful attention if you are in the audience when a speaker tells you such a thing, because it will allow you to quickly identify a fool. Anyone who tells you not to read classic papers by important scientists is someone demonstrably unworthy of giving guidance. We are not asking you to become a historian of science. We are expecting you to know where the field has come from, where it is going and who along the way provides useful reference points in the challenge to understand community ecology.

Some readers will certainly want us to say more about potential advances. This is an entirely reasonable request, since presumably we have thought a good deal about the topic while writing this book. We will therefore turn to this in a final speculative chapter. We wish to remind younger readers to first focus on mastering the basics. If you have followed our story up to and including this chapter, you have already accomplished a great deal and need not feel obliged to read further.

KEY POINTS OF THE CHAPTER

- The CATS model uses average trait values of both species and communities in a system of linear equations to predict the relative abundance of each species in a community **C** from the regional species pool **P**. This model has been tested in a variety of habitats globally.

- The CATS model chooses the distribution of species relative abundances with the maximum entropy (or evenness). This choice makes many ecologists uncomfortable, given that distributions of relative abundances in nature are rarely ever even. However, this choice is based on the logic of probabilities and is mostly statistical in nature. In fact, the linear constraints do most of the work, and the predictions are typically not even at all because they are often dominated by the few species in the pool that exhibit traits that are closest to the optimum trait value.
- There is an asymptote to the number of useful traits (between four and eight) in predictive models. The asymptote is determined by the strength of covariation among the traits of the species pool. This implies that including too many correlated traits leads to diminishing returns in predictive power.
- The Traitspace model uses Bayes' theorem to combine trait–environment relationships and multidimensional trait distributions for each species to predict the relative abundance of each species in a community **C** from the regional species pool **P**.
- Both models can be used to determine the importance of traits in community assembly. We illustrate both models using the same ephemeral wetland community from New Zealand and provide data and R code for these analyses at https://github.com/danielLaughlin/CommunityEcology.
- The Traitspace model can be modified to predict community assembly when multiple trait optima are likely to exist in a habitat.
- Each model has costs and benefits, which future work will clarify. Other alternative models may be built using these as foundations, or new approaches could be found.

8   Prospects and Possibilities

A logical structure for community ecology. Two more proposi-
tions. When is a community fully assembled? Functional groups
revisited. The canonical model and its implications. Rare spe-
cies in community ecology. Response rules and inertia in com-
munity ecology. Filters revisited. Restoration in community
ecology.

LOGICAL STRUCTURE FOR COMMUNITY ECOLOGY

The primary purpose of this book has been to introduce you to community ecology. We have proposed that community ecology is not a haphazard collection of studies falling somewhere in between population ecology and ecosystem ecology, but rather a well-defined field of inquiry, with its own logical structure and well-defined goal. We have therefore covered six main topics.

1. We have provided a unifying framework which we suggest defines community ecology: the unifying logical framework of pools, filters, traits and models for community assembly.
2. We have emphasized the need for good raw material, particularly well-defined species pools and thoughtfully prepared trait matrices.
3. We have shown how environmental factors act as filters upon species pools, and illustrated the process with several examples including drought and flooding. We have emphasized the need for better quantification of the relative power of filters in communities, and the need to prioritize important (powerful) filters in ecological studies.
4. We have provided a narrative of how the field has developed over the past century, and some of the key people who participated, and shared some of their cautionary messages.

5. We have introduced the models that currently provide tools to predict the assembly of communities from pools of species using traits.
6. Along the way, we have provided some examples of how these ideas are interwoven with the protection of natural areas, although we have resisted the temptation to provide a separate chapter on applications. In many regards, theory and applications should be inseparable.

We will now allow ourselves a few pages looking at prospects and possibilities for future advancement. These comments are primarily aimed at those who are actively working in this field, and looking into the future for key areas that need further work. We identify four areas that we consider particularly rich with opportunity: (1) species, functional groups and community assembly; (2) the canonical model for commonness and rarity – lessons for community assembly; (3) response rules and inertia in community assembly (time dependence) and (4) the role of causal factors in ecological communities.

SPECIES, FUNCTIONAL GROUPS AND COMMUNITY ASSEMBLY

Both CATS and Traitspace assemble communities from a trait matrix based on species nomenclature. This may have certain limitations. For example, in some communities information on species may be limited. There may not be time to build an entire trait matrix. Also, for purely theoretical reasons, it seems somewhat unsatisfactory to use species without considering the functional groups to which they belong (although, in fairness, the trait matrix does implicitly include information on functional groups). We have suggested in Chapter 6 that the biological diversity in communities falls into two rather different categories: diversity among functional types and redundancy within functional types. Might there be advantages to making this distinction within models of community assembly?

Let us explore a few possibilities. In order to do so, let us actually begin with a fundamental topic that we have thus far avoided: how do we know when a community is fully assembled? Getting back to the analogy of assembling a bookcase, how do we know when the bookcase is finished? Is it because it has become functional, that is, there are shelves on which we can place our books? Is it because we have run out of parts that came in the package? Is it because it looks like other bookcases we have seen elsewhere? It is possible that a community might be only partially assembled, like a bookcase that has only a couple of shelves, although the assembly guide tells us there should be six shelves. Yet, the two shelves can clearly hold at least some books.

So, how do we know when a community is assembled? We offer two propositions.

Proposition 5. A community is *functionally assembled* when there is at least one species representing each functional type that is adapted to the habitat.

Proposition 6. A community is *fully assembled* when each functional type has the maximum number of species that can coexist.

These propositions provide two reference points for the further discussion of how species might be assembled into communities containing one or more functional groups. You can find a summary of the six propositions in this book in Table 8.1. Now let us consider some possible implications of these propositions. Both might resonate with an analogy from chemistry, where the electrons in atoms each have multiple electron orbitals (functional groups), and each orbital is able to accommodate varying numbers of electrons (species). Of course, this is simply an analogy. Another analogy would be that a bookcase is functionally assembled when all the shelves are in place, and each shelf has at least one book on the topic that is assigned to that shelf. Most scholars naturally find the shelves in their offices accumulating books, since new books arrive faster than old books are lost or given away.

Table 8.1 *The six Keddy-Laughlin propositions for the assembly of ecological communities.*

Number	Proposition	Chapter
1	In any habitat, the power of a filter can be measured as the proportion of species that it removes from the species pool.	2
2	In any habitat, only a small number of filters is likely important.	2
3	In any region, a large number of species is likely controlled by the same filters.	2
4	Across multiple habitats, the power of a filter can be quantified by the strength of the relationship between traits and an environmental gradient.	5
5	A community is functionally assembled when there is at least one species representing each functional type that is adapted to the habitat.	8
6	A community is fully assembled when each functional type has the maximum number of species that can coexist.	8

A Possible Phase Space for Community Assembly

The final two propositions suggest that communities fall within a phase space defined by two axes, one axis for species and another axis for functional groups. At the lower extreme of both axes, the corner of the phase space (in this case, the lower left of Figure 8.1), we can imagine a hypothetical community with one functional type containing one species. Extending away from this end point in one dimension (vertically in Figure 8.1), we encounter a series of communities that still possess just one functional type, but contain added numbers of species. This axis of the continuum ends when the number of species for that functional type has reached a maximum ($S_{\max}$ in

Figure 8.1). It may be at a maximum for two quite different reasons. Perhaps the community contains all the species representing that functional type that occur in the species pool. Or, perhaps the species within that functional type have reached some sort of equilibrium between the rate of immigration from the pool and the rate of local extinction. In either case, in this region of the continuum, there is nearly complete convergence in traits, and maximal redundancy.

Along a second axis lie many more kinds of communities with increasing numbers of functional types. At one extreme along this axis we can imagine a community (A in Figure 8.1) with the maximum possible number of functional types (F_{max}), each functional type being represented by a single species. In this case, there is minimum convergence and minimum redundancy. Other possibilities can be imagined, involving communities with varying combinations of functional

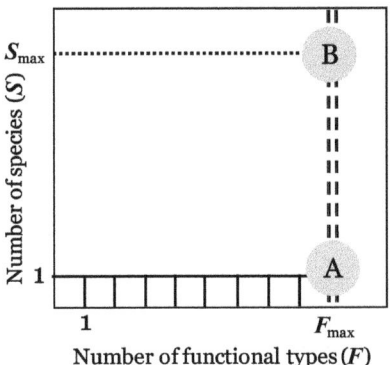

FIGURE 8.1 A community is "functionally assembled" when there is at least one species present for each functional type (point A), in other words, when $F = F_{max}$ and $S = 1$. The vertical dotted line indicates that environmental filters have set the upper limit for the number of functional types that can occur in this habitat. The community is "fully assembled" when each functional type is saturated with species (point B), in other words, when $F = F_{max}$ and $S = S_{max}$. This occurs when the immigration rate equals the emigration rate and when functional redundancy is maximized. In this figure, for simplicity, we assume each functional type has the same number of species at saturation (dotted line, S_{max}), but there is no *a priori* reason to expect this to be true.

types and numbers of species representing that functional type. Typically, most communities will contain a small number of functional types (usually fewer than 10, according to Chapter 6) and a much larger number of species. Therefore, typically $S \gg F$. In the extreme case, where all functional types are present and each functional type has reached its own S_{max}, the community would be considered "fully assembled." In this case, returning to furniture, our bookcase has all the shelves properly in place, and each bookshelf contains all the books that it can hold. Others have to be stored elsewhere in the book pool, but remain candidates to be added if a book is lost or superseded. Returning to the general problem of furniture assembly, anyone one who has actually done this kind of project knows that this process can be untidy. Frequently we are left with parts, such as extra screws, and, frequently, we find we have extra books that cannot be accommodated on the shelves. Reality is untidy.

Filters, Assembly and Coexistence

Now let us consider some of the biology that might lie behind the two propositions. The number and kind of functional types that occur in a community are determined in part by the filters operating in a particular habitat. These filters determine which subset of the species in the pool can potentially occupy the habitat. This subset contains both a set of different functional types and the possibility of multiple similar species within each functional type. For example, a pool of herbaceous wetland plants will contain both matrix and interstitial species, with multiple species representing each of the two kinds (recall Figure 6.3).

The more filters that are operating, the fewer the number of species that can occupy a habitat. That is, filters are additive, if not multiplicative, in their impact. In practice, it takes only a few filters in combination to remove the majority of the species in the pool. For example, flooding with saltwater, combined with frost, eliminates all woody species from coastal marshes (Chapter 2).

The combination of several filters selects for a small set of traits, producing ecomorphological convergence. However, it appears that while there is convergence, there can be several combinations of traits that have high enough fitness to survive those particular filters. Each combination of traits that can survive a set of filters is known as a functional type. We have seen many examples of habitats with strong filters and multiple functional types, including deserts (Figure 2.1), Mediterranean landscapes (Figure 5.5) and marshes (Figure 6.3).

The minimum number of species in a functionally assembled community is, by definition, equivalent to the number of functional types. At this point, most of the available niche space is now occupied by at least one species. We will resist the temptation to write a review of niche space and ecomorphological space, except to acknowledge the original work (Hutchinson 1958) and to point to Figure 4.7, which illustrates the possible key axes of Hutchinson's n-dimensional hypervolume in plants. The various functional groups displayed in Chapter 6 can each be seen as an attempt to map out this space, not by measuring environments directly, but rather by documenting the combinations of traits that occur in a species pool.

The number of functional types has an upper limit in any community (F_{max} in Figure 8.1). This, however, does not set an upper limit on the number of species that can occupy that community. In practice, each functional type can hold many similar, or even nearly identical, species. The presence of multiple species within the same functional type produces what we call redundancy.

Assembly, Redundancy and Coexistence

What do our two propositions mean for the vast topic of species coexistence? The coexistence of multiple similar species is often presented as a central unsolved problem in ecology. How is coexistence possible? Why does one species not exclude the other? A classic answer is that this occurs because intraspecific competition is greater than interspecific competition, an explanation rooted in analysis of the Lotka–Volterra equations (see, e.g., MacArthur 1972). A recent meta-analysis demonstrated that intraspecific competition among species pairs is 4–5-fold stronger than interspecific competition (Adler et al. 2018). Although many writers assume this Lotka–Volterra view of the world to be self-evident, there are opposing views. In fact, this conclusion about intra- and interspecific competition rather depends upon how one measures competition within a community, and how many species one considers. When we look at large numbers of species in large multispecies experiments, competitive hierarchies arise, with a few dominant species supressing large numbers of subordinate species, and in this situation most competitive interactions are not similar but asymmetric (e.g., Keddy and Shipley 1989, Keddy et al. 2000, Keddy 2001, chapter 5). In these cases, interspecific competition from the dominants is an important force affecting most other subordinate species. This view of reality is found in several models relating plant diversity to competition in communities (Grime 1973b, 1979, Keddy 1990b, Wisheu and Keddy 1992). Here, weaker competitors are excluded by interspecific competition from dominant species, unless there are countervailing forces such as grazing or fire that reduce the populations and biomass of the competitive dominants (Grime 1973, 1979, Connell 1978, Huston 1979). In spite of the many kinds of evidence across a wide array of habitats (including, as it says in Connell's title, "tropical rain forests and coral reefs") the assumptions that species interact only in pairs and that intraspecific competition must be greater than interspecific competition continue to persist. Here we wish to draw attention to this alternative view that sees a world in which asymmetric competition is respected: a world in which there are competitive hierarchies and in which recurring natural disturbances provide the primary mechanism for coexistence.

In this world, there is a further, rarely appreciated corollary about similarity and coexistence. If species are relatively similar in competitive ability, there is no obvious way for exclusion to occur at all, since intraspecific competition nearly balances interspecific competition. In the limiting case, species are so similar to one another that the distinction between intraspecific and interspecific becomes irrelevant because of the amount of phenotypic variation within each species, particularly among age classes and body sizes. Consider tropical forests, where a tree can range in size from a seedling to a canopy tree, often with corresponding changes in leaf morphology as well. Hence, one can argue that in some, perhaps many, cases there is no strong mechanism for competitive exclusion among similar species. And, more importantly, if it does exist, it is less important than the strong competitive effects from the species at the top of the hierarchy. Overall, in this context, the more similar species are, the greater the probability that they will coexist nearly indefinitely. Coexistence in this manner (we might call it competitive balance) becomes even more likely when we allow, as we surely must, small changes in the environment, and also small disturbances that remove individuals from time to time. The world is not deterministic: both fluctuations and disturbance will extend coexistence. From this perspective, the coexistence of similar species is not surprising, particularly in a world with frequent natural disturbance (Connell 1978, Grime 1979, Huston 1979). We have emphasized this alternative view of reality because the narrow and deterministic view that coexistence absolutely requires that intraspecific competition must exceed interspecific is simply not true, particularly in a world with constant natural disturbance. Indeed, one might argue that the importance of perturbations for coexistence can be traced all the way back to Hutchinson's paradox of the plankton, where seasonal change in lake conditions is offered as one reason for the (non-equilibrium) coexistence of multiple species of plankton (Hutchinson 1961). We do not deny that in certain specific limiting cases the Lotka–Volterra view that intraspecific competition must exceed interspecific competition may occasionally occur, we simply note that the vast majority of evidence – even decades ago (Keddy and Shipley 1989, Keddy 2001) – shows that there are dominant species suppressing subordinate species. And, we also draw attention to the fact that there is a strong counterargument: in the limiting case of competition among several similar species, the very similarity of the species means that small amounts of phenotypic variation or environmental change may be enough to produce coexistence.

We will resist the temptation to further review the vast literature on competitive dominance and disturbance-mediated species coexistence that has arisen since the overviews provided by Connell, Grime and Huston. That would be a book in itself. It is noteworthy that all three authors wrote about this topic at similar times, although often using quite different examples.

Returning now to our two propositions, and Figure 8.1, we observe that there are two separate factors that affect coexistence and local diversity: the number of functional types and species redundancy within functional types. We suggest that some, if not much, of the confusion about coexistence may arise out of failure to distinguish between coexistence of different functional types (coexistence of rather different species) as opposed to coexistence of similar species within one life form (coexistence of rather similar species). In general, it seems likely that the former occurs because the species are exploiting different regions of multidimensional niche space (maximizing niche volume in hyperspace while possibly minimizing niche overlap), whereas the latter occurs when species converge to exploit nearly identical regions of niche space (with maximal niche overlap). These are likely two different phenomena. It may be worth repeating that in the latter case, intraspecific competition and interspecific competition are so similar that traditional methods of thinking about competition (e.g., the Lotka–Volterra equations) may be misleading. Rather, a large number of similar species can coexist, so long as there are factors in the landscape that suppress the dominant member of a functional group. Our survey of filters in Chapter 2 shows that landscapes are full of such factors: fires, floods and intense grazing. Indeed, perhaps this explains the puzzling contradiction or paradox in Chapter 7, where the CATS model provides good prediction without explicitly including competition.

We will have to resist the temptation to launch into a discussion of "species packing," another field with origins that can be traced back more than 50 years to work including MacArthur (1958) and Hutchinson (1959). In several places we have explored topics that impinge upon species packing, such as the number of dimensions for traits (recall Figure 4.7), the variety of functional types in ecological communities (Chapter 6) and the upper limits to community assembly (Figure 8.1). The whole idea of "disturbance-mediated coexistence" also has implications for theories about species packing. However, we will leave these topics for others to explore.

The CATS Model, Competition and Coexistence: A Field Example

By answering the question about coexistence, we have set the stage for revisiting a troubling question left over from Chapter 7. Why does the CATS model work when it mostly ignores competition? One answer is that competition between similar species is, contrary to decades of dogma, really rather unimportant. Most species are far more affected by competition from dominant species, and most of them survive because natural disturbances regularly remove these dominants. Thus, the fact that we can mostly ignore intraspecific competition among similar species in the CATS model simply reflects the reality of the biological world. Competition among similar species may seem obvious in principle, and justified by Lotka–Volterra-style thinking, but as Connell, Huston, Grime and many others have convincingly shown, most of the world's ecological communities are driven by dominance and natural disturbance.

To put it another way, yes competition is important in community assembly, but not in the way that limiting similarity leads people to think. It is important in controlling the relative abundance of species within competitive hierarchies, and it may be important in interactions among functional groups, particularly in cases where sessile species are in competition for space. Within a functional type, among species that are quite similar, competition occurs, but its sorting effects are overwhelmed by other factors including direct physical factors (fire, grazing) and interspecific competition from species that are already abundant within the community. Viewed in this way, the CATS model works because species occurrences and their relative abundances are indeed mostly controlled by filters and coarse-scale competitive interactions.

Let us consider a real-life example of this phenomenon within an applied conservation context of commonness and rarity in ecological communities. So, while we are discussing competition and coexistence, let us illustrate how our view of coexistence applies to real plant communities, and how it affects the conservation management of a set of rare and threatened plant species. Our example comes from a biodiversity hotspot within southwestern Nova Scotia, where shoreline wetlands (wet meadows) have large numbers of species (Wisheu and Keddy 1989, Wisheu et al. 1994). Why do such small areas have large numbers of rare species? Does competition tell us something about the coexistence of these rare species? Well, yes and no. It depends upon the kind of competition. Many of these plants fall into one functional group of comparatively small species with evergreen rosette morphology, small seeds, low relative growth rates and weak competitive ability. Multiple experiments have shown that rare plants in this functional group (e.g., *Sabatia kennedyana*, see Figure 8.6) are rare not because of competition from other similar species (i.e., species that are ecomorphologically similar, or in the same functional group) such as *Eriocaulon septangulare* or *Coreopsis rosea* or *Hydrocotyle umbellata*. Of course, you might be able to force them to compete if you grew them in pairwise experiments in small pots, the traditional diallele experiment. Rather, all are rare because of overwhelming asymmetric competition from competitive dominants of different life forms: grasses, sedges and shrubs. Most shorelines in these regions are dominated by these life forms (Wisheu and Keddy 1994). The few locations with large numbers of rare species occur when environmental factors like summer water level fluctuations, winter ice scour and waves remove shrubs grasses and sedges, leaving herbaceous meadows on gently sloping shorelines. That is, a small set of filters creates a disturbance that removes the competitive dominants, just as Connell, Grime and Huston describe. Yes, it is possible that there is some pairwise competition among the associated rare plants, but even if it does occur, it is irrelevant to coexistence, since this habitat arises out of natural disturbances and repeated perturbations. This group of rare plants therefore illustrates an important general principle: while competition may be widespread in nature, most kinds of competition that occur are not measured by ecologists because of the predominance of pairwise thinking (Keddy 2001). We could even stretch the point to suggest that this attitudinal problem exemplifies the point made by Rigler – important causal factors in communities are frequently overlooked because people narrow their view to pairwise interactions, in which case the Lotka–Volterra model widely shared by zoologists like MacArthur (1972) in fact has led to a good deal of confusion. Instead of rose-coloured glasses, too many ecologists are wearing pairwise glasses, although the recent move to model the competitive milieu of multispecies communities is a step forward and away from this narrow view (Mayfield and Stouffer 2017, Saavedra et al. 2017). Many species may be rare not because of competition with similar species, but rather because their habitats are rare because the filters that create those habitats occur infrequently within a landscape. Which takes us back to the point that within a functional group, large numbers of similar species may indeed co-occur – just like the plankton in lakes, and trees in tropical forests, and herbs in limestone grasslands, and the corals on reefs (Connell 1978, Grime 1979, Huston 1979).

How Many Species Can a Community Hold?

To return to community assembly, a community is termed *fully assembled* when each functional type has the maximum number of species that the habitat can support. What sets this upper limit? Likely, this maximum number of species within each functional type occurs when the rate of arrival/immigration of new species is balanced by the rate of local extinction. That is, each functional group in a patch of habitat may be considered to be an island of sorts.

In some situations, the community may be able to hold additional species that have not yet dispersed to the habitat. In other situations, the number of species may have reached an equilibrium with the pool. One cannot tell without more work. One line of evidence is to compare the species found in the community with the candidate species in the pool. Are they the same? A more pragmatic way to accomplish the same goals is to compare a large number of similar locations, say alvars, and ask whether certain alvars are missing species that occur in other alvars (recall our discussion of the different ways one can assemble species pools back in Chapter 3). It is also possible to experimentally test whether the habitat is saturated by experimentally adding propagules. A considerable number of such experiments have been done, as we described in Chapter 3, and they suggest that, in general, adding new species will frequently increase richness (Myers and Harms 2009).

It is entirely possible that in some, even many, cases the species pool is simply too small to saturate each functional type. In this case, a local habitat (and community) may contain all the species that the pool provides, but still an insufficient number for the functional types to be saturated by species. Here, the lack of diversity within functional types may lie in the lack of available species with the appropriate traits, an evolutionary and/or biogeographic deficit. It might even be possible that these kinds of habitats and communities are particularly at risk of invasive species, since the habitat is so far below saturation that there are vacancies. Finally, it is entirely possible to have a community in which some functional types are saturated, and others are not.

Functional Groups Applied to the Kettlehole Wetland Example in Chapter 7

The shift in focus from species to functional groups can offer considerable advantages for prediction. The kettlehole wetland in New

Zealand, for example, contains 44 herbaceous species. As discussed in the previous chapter, accurately assigning probabilities to 44 species is difficult. However, since some species have similar traits to other species, we may treat these groups of similar species as being functionally redundant. If we were to use traits to sort each species into a smaller number of functional groups, then the problem of prediction is greatly simplified. Rather than assigning probabilities to species, we can assign probabilities to functional groups. Let's try it.

Recall that root aerenchyma was the most important trait that determined where species were found along the hydrologic gradient in this wetland (Purcell et al. 2019). We classified these 44 species into 6 groups using the "hclust" function in the R package "vegan" (Oksanen et al. 2019). Then we used the CATS model to predict the relative abundance of those six groups.

It turns out that when we reduce the complexity of the problem, our predictive power is improved. When we applied CATS to 44 species, the model explained 29 percent of the variation in observed abundances. However, when we applied CATS to six functional groups, the model explained 45 percent of the variation in observed abundances (Figure 8.2). This example is extremely simplified for the purpose of illustration – we used only a single trait to classify species into functional groups. We expect ecologists could use multiple traits to refine their classifications or use other categorical classifications (recall Chapter 6).

One Further Consideration for Functional Groups and Community Assembly: Invasive Species

Some invasive species may successfully invade a habitat because they represent a new functional type that previously was not present in the existing pool. In other cases, however, an invasive species may represent a functional type already present in the community. In this latter situation, species invade because they are similar enough to species within that functional type, and that

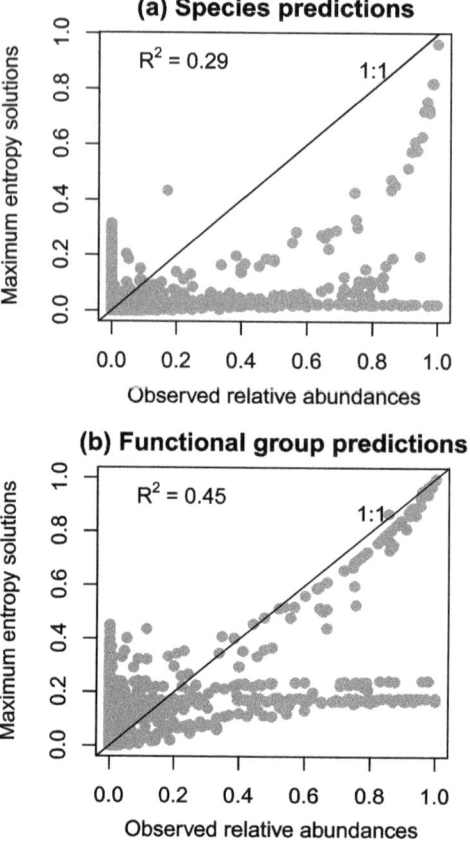

FIGURE 8.2 Comparison of CATS model predictions in the kettlehole wetland of New Zealand using (a) species and (b) functional groups. Here we illustrate the relationship between the observed relative abundances, on the x-axes, and the CATS model predictions, on the y-axes. Each dot represents a species (or functional group) in each plot. (The R code to perform the classification and subsequent CATS analysis is found in our GitHub repository.)

functional type is not yet saturated. Or, it is saturated but the species has a slight difference that allows it to replace similar native species. These offer two quite different views of how invasive species become invasive. Both mechanisms may be occurring simultaneously. Let us consider some examples.

Lythrum salicaria invaded wetlands of North America in the previous century (Thompson et al. 1987). It is an interstitial species that occupies newly created gaps. Similar species already occurred in North America, such as *Mimulus ringens* and *Verbena hastata*. The species therefore entered a functional group that already existed (interstitial species, gap colonizers) but had small advantages that allowed it to frequently replace the native gap colonizers. One of the most striking examples of this that Paul remembers from the 1980s was seeing where a bulldozer had driven across a wetland in a preceding year. That passage was now marked by two lines of *Lythrum salicaria* stretching across the marsh. In one experimental study, *Lythrum salicaria* was proved to be equal or better than six other native species at reducing the biomass of 48 different species of wetland plants (Keddy et al. 1998). It appears, therefore, that the species entered an existing functional group, but had superior competitive abilities to native species occupying the same habitat. In the terminology of Chapter 4, it is possible to screen species for their ability to suppress neighbours. This might be an important column to consider adding to trait matrices, particularly if invasive species occur in the species pool.

Phragmites australis (ssp. *australis*) has also invaded wetlands of North America (Blossey et al. 2002). Unlike *L. salicaria*, it is a clonal dominant (recall Figure 6.3). Similar species already existed in North America, such as *Schoenoplectus (Scirpus) lacustris* and *Typha latifolia*. However, *P. australis* is particularly tall and is able to exclude many other clonal species. The clones are also dense enough that they have few spaces for interstitial species. Thus, *P. australis* has a strong negative effect on plant diversity. It not only displaces morphologically similar species, but simultaneously suppresses other functional types.

These two examples, *L. salicaria* and *P. australis*, illustrate another important point in invasive ecology from the perspective of community assembly. Both of these species represented existing functional types, and so might be expected to coexist with other species already present in that functional type. However, each of them had a slight advantage over native species, mostly because of slightly higher growth rates and slightly stronger competitive abilities. Thus, they were able to displace native species within a functional type.

Sticking with wetlands, might there be examples of entirely new functional types of invasive species? *Hydrocharis morsus-ranae* invaded the wetlands of North America when an employee at Agriculture Canada brought some home from a trip to Europe and put them in a fountain outside the building where he worked. It spread into the adjoining river, and now is a conspicuous feature of wetlands in eastern North America (Catling and Dore 1982, Catling et al. 1988). The rapid spread and impact of this species was likely in part due to the fact that it represented a new functional type, an entirely floating species. The wetlands it invaded previously had only small floating species in the Lemnaceae. One can interpret this successful invasion in one of two ways. A large, floating-leaved species represents a new functional type for eastern North American wetlands, and the species spread explosively because this was a region of trait hyperspace that was not yet occupied and therefore waiting to be occupied. There is an alternative interpretation. One might say that the floating functional type was already occupied in North America by the Lemnaceae, but *H. morsus-ranae* was far superior in growth rate and potential for competitive dominance, and so it replaced the native species within this functional group. Moreover, like *P. australis*, it is able to suppress species that were growing in shallow water that previously had no floating canopy (Catling et al. 1988). To continue with the theme of functional types, we note that a vast majority of wetland plants with floating leaves are also rooted in the substrate, and can therefore be considered a different functional type (recall Figure 6.13). We note that several other dangerous invasives, including Water Hyacinth (*Eichornia crassipes*) and Water Cabbage (*Pistia stratiotes*) are also free-floating. This raises the interesting question of whether invasive species might be usefully separated into two groups: those invading existing functional groups and those that are a new kind of functional group.

THE CANONICAL MODEL FOR COMMONNESS AND RARITY: LESSONS FOR COMMUNITY ASSEMBLY

A General Pattern in Ecological Communities

There are certain biological patterns that arise in nature that seem to reflect causal factors that we do not yet understand. One of the most conspicuous of these is the canonical distribution of commonness and rarity, described by Frank Preston in two papers in *Ecology* back in 1962 (Preston 1962a, 1962b).

These patterns arise, whatever methods are used to sample those communities, and they arise from mostly unknown mechanisms of community assembly. Preston's work addressed an important general rule, if not an actual law, of community ecology: all community vectors have highly unequal relative abundances described by the canonical or log-normal distribution. The observation of unequal abundances is widely discussed in community ecology, and has a rich nomenclature of its own. There is Frank Preston's own work (1962a, 1962b), where this pattern is called the canonical distribution, in which the data are displayed in different bins or shelves (octaves in his words) of abundance. We will eventually show this canonical pattern (hint, see Figure 8.5), but first want to show how this pattern in relative abundance shows up in communities as the log-normal distribution and as dominance–diversity curves. There is a huge literature on the log-normal distribution, as well as other models for relative abundance, such as MacArthur's broken stick model, and various overviews written on comparison of these models (e.g., Pielou 1975, May 1981). There is also a large literature on dominance–diversity curves (Whittaker 1965 is a classic example), which are another way of presenting relative abundance data. Dominance–diversity curves show the species in rank order from left to right, with relative abundance (often log transformed) on the vertical axis, as shown in Figure 8.3. The measure of relative abundance can range from number of individuals to biomass to cover, with the specifics often differing from one study to another. Yet another way of showing the relative abundance of species is to record the proportion of sample units in which they occur – that is, to record species frequencies. It was Raunkiaer himself (1908) who reported the law of frequency, that a *J*-shaped distribution is widespread, that is, a majority of species occur in only a few quadrats. In some habitats, like species-rich pine savannas in southeastern North America, the vast majority of species occur in less than 10 percent of the samples (Keddy et al. 2006, Clark et al. 2008). And, then, there is an entire literature on diversity itself, particularly on measures of diversity, with a focus upon how one number (a diversity index, such as Shannon's diversity index, described in Chapter 7) might be calculated to describe the degree of equality of relative abundances in a community vector or other sample (e.g., Peet 1974, Pielou 1975, Kempton 1979). Hence, without reviewing any of these topics, we can agree that in most communities a small number of species are numerically dominant. Another way of describing the canonical pattern is to say that for a given number of species, entropy and evenness tend toward a minimum.

Adding the Canonical Distribution to the CATS Model

In the previous chapter we explored the underlying theory of the CATS model. Recall that this model chooses the maximum entropy solution (Equation 7.3), that is, it seeks to maximize evenness – choosing a community vector where species are rather similar in their relative abundances. Given all that we know about relative abundance distributions in nature, wouldn't the canonical distribution, or perhaps another minimum entropy solution, be closer to reality? Let us try out this possibility, returning to a now familiar example, the kettlehole wetland where species are filtered along a flooding gradient according to their root aerenchyma (Purcell et al. 2019). What happens to our predictions of relative abundances in the kettlehole wetland

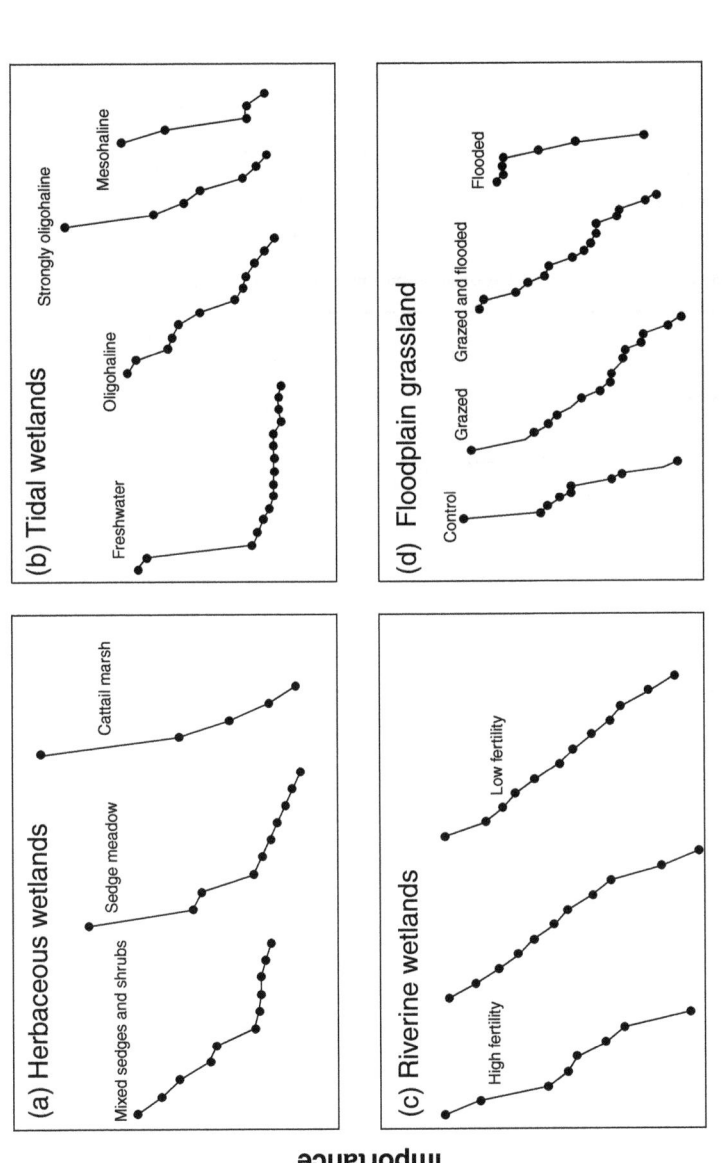

Species rank

FIGURE 8.3 Dominance–diversity curves, also known as ranked abundance lists, for 14 plant communities within four different wetlands. The measures of importance, on the vertical axis, vary among these studies: (a) freshwater herbaceous wetlands (New Jersey, log of annual net production, using data in Jervis 1969); (b) tidal wetlands (South Carolina, relative density plus relative biomass, after Latham et al. 1994); (c) riparian herbaceous wetlands (Ottawa River, log of biomass, using raw data of Day et al. 1988); (d) floodplain grassland (Argentina, log basal cover, after Chaneton and Facelli 1991).

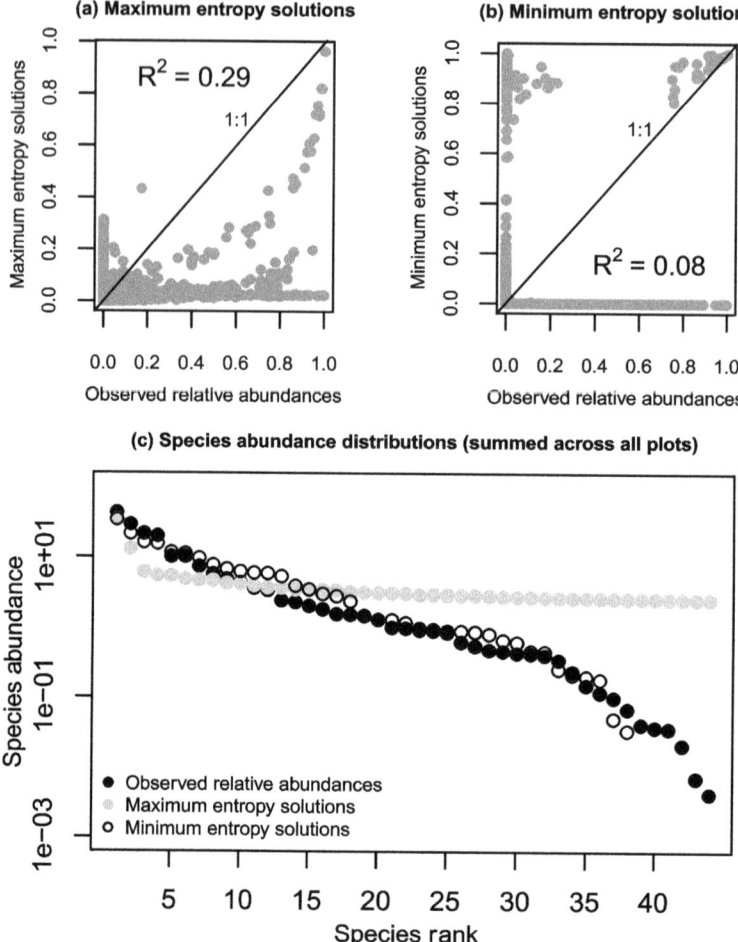

FIGURE 8.4 Comparison of results that either maximize or minimize entropy in the CATS model (recall Equation 7.3). The comparison produces what look to be paradoxical results. First, contrary to our expectations from the canonical view of reality, the maximum entropy (maximum evenness) solution (a) gives a far better fit to the observed abundances than the minimum entropy solution in (b). Second, when the abundances of species are summed across all the plots, the minimum entropy solution produces a relative abundance distribution that mimics the observed rank abundance distribution far better than the maximum entropy solution. (Data from Purcell et al. (2019).)

when we minimize entropy rather than maximize it? Figure 8.4 gives a surprising result. They get worse. Much worse.

The top panels in Figure 8.4 illustrate the fit of two models: the first maximizes entropy and the second minimizes entropy. Recall that entropy is equivalent to evenness, so by maximizing entropy we maximize the evenness of the relative abundance distribution. That is, we reduce the degree of dominance by a few species. The overall fit of the maximum entropy model is much better (R^2 = 0.29) than the minimum entropy model (R^2 = 0.08). There are a few species that are better predicted by minimizing entropy (see top right corner of Figure 8.4b), but the majority of the predictions are not good at all because most occur along the edges of Figure 8.4b. We return to this matter shortly.

We can also ask how the predicted relative abundance distributions in the resulting communities compare to the observed relative abundance distributions. So, we next summed the abundances within species across all the plots and plotted the observed, maximum entropy and minimum entropy relative abundance distributions (Figure 8.4c). The minimum entropy solutions were much closer to the observed canonical distribution, whereas the maximum entropy solution produced a far more even distribution of abundances! Are these results contradictory? No, they are not. A relative abundance distribution is nameless. The species identities are removed and the only important property of the species that is left is the rank order. In other words, despite the fact that the overall patterns look similar, the actual rank order of species differs greatly between the observed and minimum entropy solutions. To put it another way, the minimum entropy (canonical) solution picked dominant species that were more often less correct than the true dominant species.

Still, something is puzzling. We have just established that the canonical distribution is one of the most widespread general patterns in ecology. We have already mentioned that the CATS criterion of maximum evenness therefore seems counter to ecological theory. Yet, when we replace the maximum evenness criterion with the canonical, the predictions are noticeably worse! Since this is the last chapter of the book, we will mostly have to leave this apparent paradox as an open question in community assembly. As a first step, we need to come up with some working hypotheses for this strange result. Three possibilities occur to us.

First, we know that filters select for a small set of traits, and therefore are a force for convergence in community assembly from pools. What sets an upper limit for convergence? One possibility is competition, which takes us all the way back to the early chapters of this book. If limiting similarity from competition occurs, then we have a force pushing back against convergence. The actual community is therefore a result of two opposing forces: trait convergence driven by filtering, and trait divergence resulting from competition. We noted in Chapter 7 that the CATS model ignores competition. So, here is a possible resolution. The criterion for maximum evenness is in fact a crude way of incorporating the effects of competitive pressures for trait convergence. If we look at Equation 7.3 in this light, we could speculate that CATS does include competition – by using evenness as a surrogate for limiting similarity, removing solutions in which convergence is overestimated. Problem solved.

There is a second possibility. Again, begin with the view that filters are selecting for maximum convergence. This should lead to only a few species dominating the community. These are the few species that have the best traits to resist the effects of those filters. We have just considered that intraspecific competition might be a counter-force to this dominance. Can we imagine any other counter-forces to high degrees of dominance? Yes, we can. We have also just discussed how natural disturbances tend to reduce dominance and increase diversity. We have described how Connell, Grime and Huston have all made the case for how diversity is enhanced by natural disturbance. So, here is another possible resolution. The criterion for maximum evenness is in fact a crude way of incorporating the effects of ongoing natural disturbances that prevent full trait convergence. It is a way of forcing an outcome that includes species that do not quite fit the preferred set of traits, but nonetheless can persist in similar conditions. We have just given the example of coastal plain plants in Nova Scotia. These species can survive the key filters (flooding, infertility, winter ice cover), but they are normally excluded from wetlands because they cannot compete with the dominant life forms that normally occupy this habitat: grasses, sedges and shrubs. So, thinking in this way, the criterion for maximum evenness might be interpreted as a tool for ensuring that some species that can tolerate the filters, but cannot tolerate the competitive dominants, will still occur in the community vector, albeit at very low levels of abundance (among the smallest of the p_is in the output vector). Without Equation 7.3, the CATS model would not include such minor species in the output vector.

Using the same kind of arguments, we could imagine a third possibility. Are there any other factors that provide a counter-force to dominance by a few species? Yes, a third might be environmental heterogeneity: species that occur in rarer habitats, like vernal pools, rock outcrops, cracks in limestone pavement, animal burrows or dung piles. We could interpret the maximum evenness criterion as forcing the output to include local environmental factors that provide opportunities to escape from the sheer physical forcing of the environmental filters.

All three of these suggestions assume that the CATS model tends to overestimate the effects of convergence, and that the evenness criterion ensures that other species occur, albeit at rather low levels of abundance. We could think of Equation 7.3 as ensuring the realistic outcome that nearly all community vectors will include a lower tail of species that are rare.

So this is our list of speculative reasons for the results in Figure 8.4. The first one restores competition to its rightful place in communities, albeit by a sleight of hand. The second two postulate that in a world filled with disturbance and heterogeneity, we cannot make community outcomes entirely deterministic. Some species will find a way to exploit natural disturbances, or other nooks and crannies in the naturally heterogeneous world. The evenness criterion ensures that rare species are included in the output from the model. Which, conveniently, allows us to say more about the lower tail, the rare species.

The Lower Tail of the Canonical Distribution, and Tail Trimming

We have so far emphasized the rule that a few species will dominate any community vector. It is the dominants that we usually see, and sample, in describing ecological communities. Now let us turn that observation around and consider the reverse of this statement: in any community vector, a considerable number of the species will be represented by quite small amounts of biomass. This is often referred to as the lower tail of the distribution, and Preston wrote about situations in which such species are so rare that they are often not encountered, and hence exist below a "veil line" that is only removed by ever more intensive sampling. This, of course, takes us back to the issue of species accumulation curves discussed in Chapter 3 (e.g., Figure 3.10). One could argue that many ecologists already ignore this lowest tail of species abundances for practical reasons: it takes more effort to sample rare species. Moreover, these rare species increase identification problems. And, in addition, since the information on them is limited, they complicate data analysis. Thus, rare species are often excluded from multivariate studies, although frequently one has to read the methods carefully to see which species were omitted. For simplicity, we will call this practice "tail trimming." It may be an entirely legitimate practice when the goal is to predict the abundance of dominant species in ecological communities. On the other hand, it is also throwing away useful information if the goal is to understand rarity. In other cases, we suspect that rare species are often omitted not for any theoretical reasons, but simply because of the difficulty of finding field assistants who can identify rare species, particularly in groups with large numbers of species (e.g., the Cyperaceae and Poaceae).

Indeed, it would be a useful exercise for someone to read through the past 50 years of ecological studies to quantify the amount of tail trimming that is standard practice. Often, the information is not clearly given in the methods, so we cannot offer a table documenting the proportion of species that are trimmed from data in general. But we can give a selection of examples.

One guide to data analysis (McCune and Grace 2002) says "Deleting rare species is a useful way of reducing the bulk and noise in your data set without losing much information" (p. 75). "As a general rule of thumb, consider deleting species that occur in less than 5% of the sample units" (p. 75). An older and widely used guide to multivariate analysis, by Gauch (1982), also treats rare species as a potential difficulty in analysis: "Rare species are usually deleted from a data matrix prior to multivariate analysis ... Obviously rare is a relative term, but typical criteria include species occurring in less than about 5% of the samples or in fewer than about 5 to 20 of the samples" (pp. 213–214). Following such advice, here is a real example from Paul's work (this example ensures that it does not look like we are criticizing others without acknowledging our own decisions): in the study of plant communities of the Ottawa River (Day et al. 1988), eight vegetation types were produced by Twinspan after "removing species with a single occurrence, and species with <0.1 g in any cluster." Daniel admits to being equally guilty of this practice in his multivariate analyses (e.g., Laughlin et al. 2004).

Sometimes the issue of tail trimming is approached in the opposite way – by not collecting data on rare species in the first place. In the CATS example from Morocco, Frenette-Dussault et al. (2013) explicitly considered only 34 "dominant" species. Many more were obviously excluded, since this adjoins a global biodiversity hotspot. Laughlin et al. (2012) considered just nine species of trees, which was the full list of canopy tree species in their sample, but there were dozens if not hundreds of understory herbaceous and shrub species that were excluded from the analysis. In the words of Gauch (1982), "Field work is expedited if rare species are not even recorded in the first place" (p. 214). In Louisiana pine savannas (Keddy et al. 2000), it was extremely difficult to find people qualified to collect plant data, much less to accurately name the voucher specimens for the enormous number of sedges and grasses found in these savannas: what is one to do when competent field botanists have retired and been replaced by molecular biologists? Indeed, new species are still being reported from the pine savanna described in Keddy et al. (2000). With data such as that shown in Figure 8.3, throwing out the rare species could mean throwing out the majority of the species.

The general justification for such practices is that they make the analysis easier. Further, in all these examples, the resulting models incorporate a vast proportion of the biomass. So it would seem that the canonical distribution does indeed provide some justification for this kind of simplification. Yet there are dissenting voices. Even enthusiastic advocates of multivariate techniques have expressed concern about the reduced weighting given to rare species in the data set, let alone "the wholesale elimination of rare species from the data prior to analysis ... Neither approach can be condoned on theoretical or biological grounds. Instead, practitioners should seek ordination methodologies and strategies that implicitly recognize and account for the expected log-linear species abundance distribution of biotic survey data" (Kenkel 2006, p. 667).

So far, we have seen that for many reasons, and across a wide array of situations, ecologists are making the decisions to focus upon the dominant species – the ones that are most common in a landscape. Although this is often done for logistical or methodological reasons, it might be better to see all of these decisions falling into a similar category: trimming the lower tail of the canonical (or log-normal) distribution. All of which, by the way, should not be used as an excuse for hiring field teams who lack the expertise to identify rare species, nor as an excuse for simple laziness when dealing with difficult taxonomic groups!

It may seem at this point that we have been spending rather a lot of time on what is a standard practice for data analysis, and one that seems to be justified by theoretical considerations like the log-normal distribution of species abundances. So why are we using up valuable space in the last chapter?

There are at least two quite different reasons for putting so much emphasis upon this topic. First, we do have an important general principle: we can simplify the process of community assembly by focusing on the common species in landscapes. All those species that are a relatively small proportion of the biomass can conveniently be set aside ... sometimes. Keeping in mind the two tails of the canonical distribution, we might therefore consider that community assembly actually has two quite different elements: the assembly of species that will dominate the community, and the assembly of species that will remain rare (Figure 8.5).

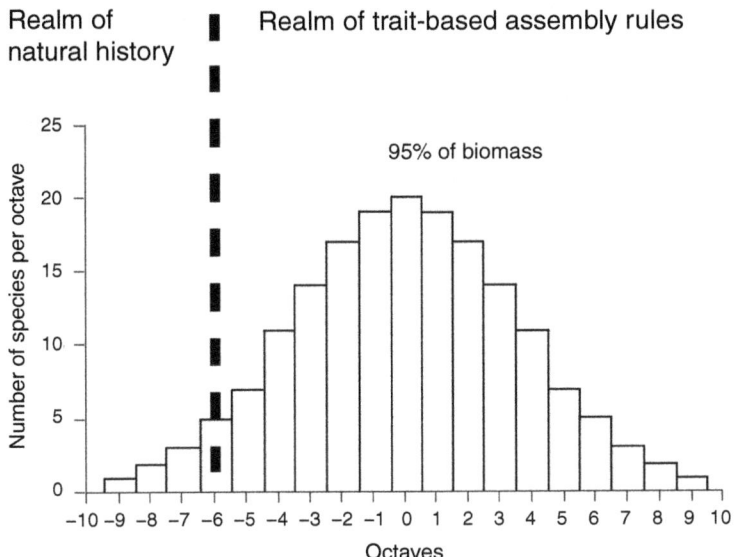

FIGURE 8.5 All ecological communities have a canonical (or, mostly lognormal) distribution of relative abundances, in which a few species are extremely abundant (far right), many species have intermediate abundance, and some are rare (far left.) When this work was first presented, the horizontal categories or bins were called octaves, so we have kept this label in this figure. Note that dominance–diversity plots (Figure 8.3) begin with the most abundant species on the far left of each line. (If you can reconcile these quite different plots of similar data in Figures 8.3 to 8.5, you can consider yourself to have a good grasp of diversity patterns within community vectors.)

Perhaps this general pattern produces two different approaches, or at least two different components within community ecology. Assembly models so far focus on the abundant species. Yes, we know that every community will also have some rare species, but the causes may be hidden within a complex mixture of natural historical details. So, we set them aside, at least in the short run. Thus, as Figure 8.5 suggests, this is one way of thinking about the difference between natural history and science.

The difference between natural history and science perhaps deserves more attention, since this is a book about science, yet it does mention natural history. Paul has put a good deal of thought into the difference between science and natural history, for many years. When he was an undergraduate, he spent the winters studying science, with a heavy emphasis upon physics and chemistry, while each summer he worked as a naturalist in Algonquin Provincial Park. Hence, early in his career he had to confront the differences in world-view. We can consider that science and natural history are two ends of a continuum in exploring reality or, equally, as two Venn diagrams with some overlap, yet different centroids. There are two sets of criteria that might be used for distinguishing between them.

First, let us think of science as being a kind of multiple regression model: science is concerned with maximizing the explanatory power of a few factors, while natural history is more concerned with the unexplained variance. In physics, there are only four forces needed to comprehend physical reality. In this book on community ecology, we have similarly suggested that only a small number of filters is needed to comprehend the kinds of communities we observe. Natural history, on the other hand, is much more focused on details. Complicated explanations for the distribution and abundance of species are accepted, and are not seen as a problem. A scientist would want to focus on the most common forest birds, and the factors that control their abundance. The naturalist seeks out the rare birds, including those that are so rarely seen their occurrence is entirely unpredictable. If we think of multiple regression as a general way of understanding the world, science focuses upon simple models with a small number of predictors, while natural history seeks models that are extremely complicated, so much so that they may be impossible. (Just when, if ever, will another Anna's Hummingbird appear in Carleton Place? (Chapter 3)). In science, we also want to avoid over-fitting, while in natural history this is not an issue. In science, the presence of the rare plant *Saxifraga aizoon* on a single cliff in Algonquin Provincial Park would be of limited interest since the plant comprises only the tiniest amount of biomass in the park, which is a landscape dominated by woody plants. In natural history, the same rare plant stands out for particular attention: the details of its location, how it persisted since the ice age, why there should be a calcium-demanding plant in a mostly gneiss landscape, when the plant flowers, what pollinators visit the flowers, and so on. The case of Bladder Fern (*Cystopteris bulbifera*) on the Keddy property presents a similar problem, just at a smaller scale (recall Chapter 3). Thus, at the risk of overgeneralizing, science is particularly concerned with simple models that predict the most common species in a landscape, the right-hand end of Figure 8.5, while natural history is particularly concerned with the unusual species in the landscape, the left-hand end of Figure 8.5. For a general scientific model of Algonquin Park, we are primarily concerned with large woody species and the role of filters like flooding, soil depth and fire in controlling their abundances. For a natural history view of Algonquin Park, the persistence of a rare calcicole from the last ice age, a minuscule species at a single location in the park, is actually a rather important story.

Another way of distinguishing between science and natural history is to consider the motivation for the research. Paul has written elsewhere about how unhappy scientists become when you discuss motivation, and has received some quite unpleasant criticism for raising this point. He thinks that this irrational criticism shows that he has correctly identified an important issue. A scientist goes out to collect information principally in order to test a hypothesis or build a predictive model. A naturalist goes out to collect information primarily for the sensual pleasure of making observations. The first seeks predictive power, the second is seeking personal pleasure. Of course, it is convenient to pretend that this distinction does not exist. In our experience, one frequently meets biologists who are fond of certain organisms, and who enjoy visiting certain parts of the world, and who think that because they have a faculty position, the mass of less educated citizens are obligated to pay them to study their favourite species and/or location. Fly fishermen want to study salmon, hunters want to study ducks, bird watchers want to study birds in the Galapagos, and so on. In other cases, people are attracted by the perquisites of power: pick-up trucks, float planes, airboats and helicopters, for example. In conferences, one can sometimes observe this phenomenon: speakers who place more attention on vehicles than upon their hypotheses. In some cases, it is apparent that for some people, personal self-interest trumps the advancement of science, except for the first or second paragraphs of the grant application. In summary, building scientific models and experiencing sensual pleasures are orthogonal axes. A fortunate scientist will find elements of both in their career. But to argue that they are one and the same does a serious disservice to science. It also wastes a good deal of money from the national treasury, since the public has the reasonable expectation that scientists are focused on solving important problems, not indulging personal pursuits. A personal example: when Paul waded through icy water in the rain with leeches trying to attach to his arms in Muskoka, he knew this was necessary to do good science. On the other hand, when he was relaxing and watching hummingbirds on the front porch of Asa Wright Nature Centre in Trinidad, he knew he was doing natural history. To return to the question of rarity, the scientist sees it as a problem to be solved (or, in some cases, as we have seen, as noise to be ignored) while the naturalist sees it as a particular source of sensual pleasure, and possible status as well. If you have spent time in the natural history community, you will know that high status is conferred by discovery of rare species, and long and intricate stories about such species are lovingly retold.

This leaves us with a rather unsatisfactory situation. What about that lower tail and all the species found there? It would be wrong to simply pretend that it does not exist or to write it off as natural history. That is not the point of the preceding paragraphs. We know (Figures 8.3–8.5) that there will be many species in that tail, but we still do not know why they are there. Perhaps, even, for each individual species in that tail there is a singular, exceptional and mostly unknown explanation for its presence and low abundance. That is to say, we can predict with confidence that some species will be rare, but when it comes to understanding why they are rare, well, that is another problem entirely. Jesus made a similar observation, in the context of income rather than biomass distributions: "the poor shall always be with us" (Matthew 26:11). In this case, however good the science of community assembly becomes, there will always be residual variance to delight field naturalists – on their holidays.

So, first the good news: the canonical distribution provides a reliable tool for simplifying predictive models of community assembly. What about costs, however? There are two reasons for being cautious with trimming the lower tail. First, this practice may misrepresent the actual communities that exist, and thereby distort our understanding of community ecology. Second, this practice may misrepresent communities in ways that explicitly affect the management of rare and endangered species.

Sample Rarity and Conservation Rarity: An Important Distinction

Thus far we have used the word "rare" in a rather general way in this section, in the context of Figure 8.3, to mean species that occur in relatively small numbers or as a relatively small proportion of biomass. In this context, a rare species is the opposite of a dominant species. In some cases, in order to avoid the use of the word rare, we have referred to them as "the lower tail." Here is one reason for being cautious about this word. "Rare" has another meaning: there are increasing numbers of species that are rare on Earth, and at risk of extinction. The loss of the world's rich array of species, from the charismatic rhinoceros to little-known insects, is an enormous problem (Ehrlich and Ehrlich 1981, Pimm 2001). It is getting worse. The loss of biodiversity is one of the great crises facing humanity. It would therefore be wrong to write a book on community ecology that does not at least mention the ongoing threats to biodiversity, or to leave readers thinking community ecologists can ignore rare species. In the conservation context, particularly in the management of parks and natural areas, "rare" species are hugely important. There are now well-defined categories of rarity, including terms such as "threatened" or "endangered," labels that are carefully assigned by various governmental and non-governmental organizations to describe degrees of risk. In this frame of reference, rare species are often species at risk of extinction. Such species need extra attention when decisions are being made about management of parks and reserves in particular, and landscapes in general. And yet, we have just explored how it is standard practice to exclude rare species from studies of ecological communities! What is going on here?

Partly, we have a language problem. The word *rare* can have two quite different meanings. So, we are going to rename them in the rest of this discussion. *Sample rarity* means simply that a species is relatively uncommon in a sample, that is, it is in the lower tail of the species abundance distribution in a particular sample, location or community. Possibly, it would be best to simply call these lower-tail species, and avoid the word rarity altogether. *Conservation rarity* means that biologists have carefully assessed the distribution and abundance of the species (and other data, such as trends in population size) in order to provide official designations of conservation rarity. The IUCN Red List is the standard list of species at risk (www .iucnredlist.org). Established back in 1964, and maintained by the International Union for Conservation of Nature, the list sadly has grown to include more than 116,000 species, including the iconic Giant Panda, the Basra Reed Warbler, the California Condor and the Chinese Giant Salamander. Overall, the species threatened with extinction include, at the time we write, "41% of amphibians, 34% of conifers, 33% of reef building corals, 25% of mammals and 14% of birds."

Let us say more about conservation rarity, since we have so far not mentioned the topic. The IUCN list is a global list. Most countries have national lists. In Canada, for example, the Committee on the Status of Endangered Wildlife in Canada (COSEWIC) meets regularly to review information on rare species in Canada, and assign them to categories including "not at risk," "special concern," "threatened," "endangered" and "extinct." Thus, it is possible that a species may be rare in Canada but not at risk. Or, it may be rare and also endangered. The point is that when dealing with *conservation rarity*, words like "rare," "threatened" or "endangered" often have explicit meanings and need to be used precisely. These words also have geographical context. A species may be threatened in Canada, but not in the adjoining USA, for example, or it may be threatened in all jurisdictions.

In Figure 8.6 we use the example of Plymouth Gentian (*Sabatia kennedyana*), a threatened wetland plant restricted to eastern North America. It was likely more common when sea levels were lower during the ice age, when it may have occurred elsewhere along the then exposed continental shelf, but the species is now restricted to small areas with infertile shorelines and pools along the east coast of North America. Some decades ago Paul actually documented the largest population of this species in Canada (Keddy 1985, Wisheu and Keddy 1989), and raised the money needed to buy a key property as Nova Scotia's first nature reserve, now called the Tusket River Nature Reserve. More recently, the Nova Scotia Nature Trust has acquired not one but two adjoining properties as the Wilsons Lake Conservation Lands, greatly expanding the protected area to more than 1,000 acres. This example illustrates a general principle: professors can be active in conservation activities even while being busy with teaching and research. Paul says that the chairman of his biology department at that time warned him, in writing, that he was spending too much time on conservation, and that it would harm his career.

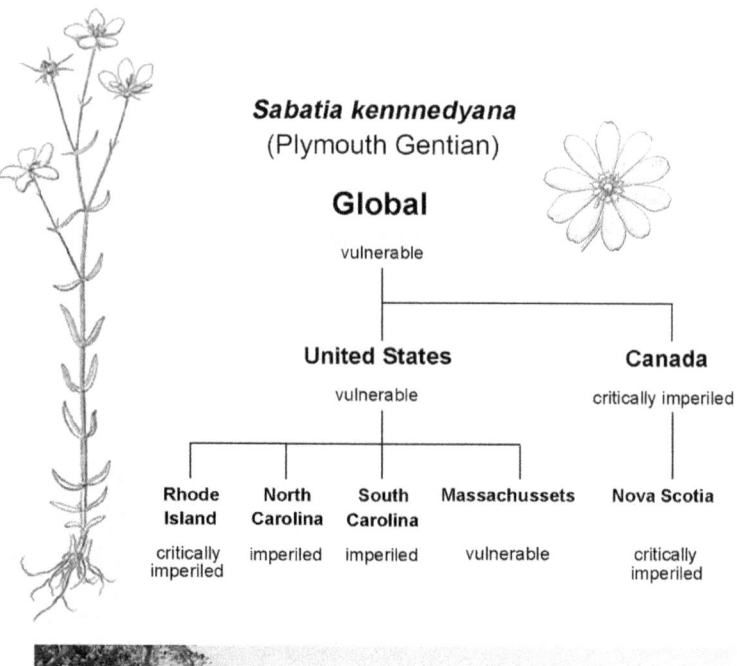

FIGURE 8.6 The status of a rare species, the Plymouth Gentian (*Sabatia kennedyana*), and some typical shoreline habitat. In this case, the terminology for conservation rarity comes from the NatureServe Explorer, NatureServe, Arlington, Virginia (https://explorer .natureserve.org). Their system includes three levels, *critically imperilled* (5 or fewer occurrences, or 1,000 or fewer individuals), *imperilled* (6–20 occurrences, or 1,001–3,000 individuals) and *vulnerable* (21–100 occurrences, or 3,001–10,000 individuals) (illustration and photograph by Cathy Keddy, line drawings by Karen Velmure).

Since this chapter is partly about the future of ecology, we think it is important to remind younger scientists that they can make a difference if they choose important goals and stick with them. We include this example not to brag but to lead by example. We invite you to imagine a future in which each professor of ecology aspires to leave behind not only well-educated students and useful research papers, but also several hundred acres of protected land.

All conservation rare species, and all national and regional lists, are nested within the IUCN Red List, which deals with ultimate survival of a species on Earth, since it transcends political boundaries. Of course, species that have very small distributions may be endangered at both a national or state scale, as well as at the international scale. It is important for students to understand how this status is assigned in their own jurisdictions, and how the conservation status of a species is nested within political entities of increasing scale.

Why have we spent so much time on nomenclature for rare and endangered species? Because the canonical distribution and the lognormal distribution do not include any such evaluations: they describe a different kind of rarity, *sample rarity*, which is simply the relatively low numbers of certain species in a particular sample. Thus, back in Figure 8.3, the species at the lower (right) tail are typically described as "rare" in each of those habitats. They are rare because not many of them were found. This pattern is an ecological law that governs how species occur in community vectors.

Wait, it gets more complicated still. Having just distinguished carefully between *sample rarity* (the lower tail of species in samples) and *conservation rarity* (assigned by conservation agencies), we now have to add a further complication. There is some overlap between data rarity and conservation rarity, and it is scale-dependent. In certain rare habitats, one can find an occasional large population of a species that is rare or threatened at the global scale. If, say, one was to visit the Wilsons Lake Nature Reserve in Nova Scotia, at the right time of year, and during a drought year, one might see hundreds of flowering plants of Plymouth

Gentian on one shoreline. It might look common here – except that there are only a handful of such sites remaining in the world.

So, returning to community models, we can see a problem with trimming the tail in our ecological models. The species at the lower end of the canonical distribution often contribute a large portion of the observed biodiversity in a habitat. And this oft-discarded tail may be the very part of the community vector that is most relevant to the management of rare and endangered species!

Let us conclude this perplexing situation by talking briefly about causes of rarity. Many, many studies over the past 50 years have described selected rare species, sometimes comparing them to similar common species, and asking if particular ecological traits (broadly defined) are a predictor of rarity. We think it is safe to say that at this point, few specific traits have emerged. There are some notable exceptions. One candidate trait is the inherent rate of reproduction. There is a simple and rather convincing argument that any species that reproduces slower than prevailing interest rates is at risk of extinction (Clark 1973), since the incentive always exists to kill the final individuals and put the resulting money in the bank for a higher rate of return. This could explain why species like tigers, tuna and rhinoceroses are still being hunted to extinction, and why the bush meat industry is emptying the world's ecosystems of animal flesh (Redford 1992). More recently, Carmona et al. (2021) have examined the life history traits of 75,000 species at risk of extinction, and concluded that extinction risk "is not randomly distributed, but localized in certain regions of trait space." The key traits for extinction risk are "large size, slow pace of life, or low fecundity." Note that the last trait is the one identified by Clark (1973). Note, too, the last two traits suggest that stress tolerators (species with inherently slow metabolism, including slow growth rates (Grime 1977, Grime and Pierce 2012) are at risk overall across a wide range of species and locations. In this regard, it may be noteworthy that *Sabatia kennedyana* in Figure 8.5 has evergreen foliage and an inherently low growth rate.

At the same time, it may also be that generalization is impossible, and, as already noted, there may be individual causes for each rare species. The last individuals of any rare species might be removed by a stochastic event like a storm or a highway or a bulldozer, although larger and more predictable forces may have created the precarious situation in the first place.

It also appears that there are certain habitats that support large numbers of rare species, and hence, landscapes have biodiversity hotspots (Myers et al. 2000). Although the concept of biodiversity hotspots is mostly applied at the global scale, there seems to be a tendency for this to occur at smaller scales. The Wilsons Lake wetlands mentioned above, which provide habitat for *Sabatia kennedyana* (Figure 8.6), have many other rare and threatened species (Wisheu and Keddy 1989), which makes it a biodiversity hotspot within the province of Nova Scotia.

In 1981 Deborah Rabinowitz tried to bring some order to this perplexing field by postulating that there are at least seven different kinds of rarity (Table 8.2). In this work, she was mostly describing sample rarity at relatively large scales, which converges on the kind of conservation rarity considered by groups like COSEWIC and the IUCN. We will have to leave it up to you to read her book chapter, and the vast number of studies that have followed. Three approaches seem to be emerging. First, it may be possible to use estimates of population growth rates to predict risk of extinction (Clark 1973). Second, it may yet be possible to screen large numbers of species for traits associated with risk of extinction (Carmona et al. 2021). Third, it is possible to assess habitats at risk, beginning with maps of biodiversity hotspots at global and regional scales (Myers et al. 2000). It may also be useful to distinguish between species that naturally occur at the lower end of the abundance distribution, and those that have been driven there by humans – perhaps there are important differences between the two.

Putting Rare Species into Community Models

There is now a huge species-specific literature dealing with rare species. There is even a log-normal distribution of such studies, with some charismatic species having huge numbers of studies and attracting great conservation effort (e.g., rhinoceroses, pandas, gorillas, whales,

Table 8.2 *Seven potential kinds of rarity based upon three characteristics of the species: their geographic range, their habitat specificity and their local population size (Rabinowitz 1981). Note that Rabinowitz is not addressing sample rarity, at least in single samples, but rarity in landscapes. Rabinowitz emphasizes that this table addresses patterns in species occurrence, not causation. This kind of rarity may also be conservation rarity, but it is conceivable that species may be rare but not at risk.*

Geographic range	Large		Small	
Habitat specificity	Wide	Narrow	Wide	Narrow
Local population size				
Large, dominant somewhere	(Not rare)	Locally abundant over a large range in a specific habitat	Locally abundant in several habitats but restricted geographically	Locally abundant in a specific habitat but restricted geographically
Small, non-dominant	Constantly sparse over a large range and in several habitats	Constantly sparse in a specific habitat but over a large range	Constantly sparse and geographically restricted in several habitats	Constantly sparse and geographically restricted in a specific habitat

condors) while other species decline into extinction in comparative obscurity. How might this body of work be fitted into our framework for community ecology? Looking ahead, one option, rooted in natural history (*sensu* Figure 8.5) may involve scientists producing more species-specific models. But there are already more than 100,000 species on the IUCN Red List, and that would be a lot of models.

We may therefore, in the intermediate term, need to break a community into two groups of species. For the common (or dominant) species, community assembly models may be used to focus on predicting species composition for the majority of the biomass (say, 95 percent). At this scale, one might reasonably ignore a significant number of locally rare species, focusing on general rules about traits and environments. Then, there may be a line below which general principles do not apply. A general model for ecological assembly for Algonquin Park would be expected to predict the abundance of trees, not the occurrence of *Saxifraga aizoon* on an isolated cliff. A general model for ecological assembly on the Keddy property would be expected to correctly predict oak and ironwood dominance on rock ridges, but not the occurrence of *Cystopteris bulbifera* on a marble boulder. Hence, we return to that vertical line shown in Figure 8.5, separating the canonical distribution into species likely to be explained by trait-based assembly rules, and those for which more (perhaps much more) detailed information may be needed. At the very least, this line may indicate the need for added causal factors and/or more traits, taking us into that risky realm that can end up with overfitting and a complicated natural history tale for every observation.

Whether it is possible, or even desirable, to reach a point where we have a large number of single-species models is a topic for consideration. How many models is enough? And where do we draw the line (Figure 8.5)? Again, we hear the voice of Rigler and consider the enormous number of possible pairwise interactions in ecological communities – not to mention higher-order interactions. And we also know from theoreticians that large systems of differential equations for single species, when combined, can eventually simply produce chaos (May 1973).

And, similarly, even simple equations require so much biological information (e.g., competition coefficients, growth rates) and so many further simplifications (e.g., ignoring higher-order interactions) that they cannot be reliably used to predict community responses (Shipley 2010). Therefore, we suggest that while it is important that community ecology take rare species into account, for both theoretical and applied reasons, it is not yet clear how best to accomplish this task. It is a problem that invites a solution.

Regardless of this discussion of science and natural history, the immediate practical application of the canonical distribution remains clear: in order to protect the rare species of this world, it is still a first-order priority to protect the wild landscapes in which they live. The more land that is set aside, the greater the likelihood that rare species will survive into the future. This is one rule for conservation that we can state with certainty. Let us repeat: the greater the area of protected land, the greater the number of protected species. That is rooted in both common sense, as well as in the canonical distribution, as well as in the principles of island biogeography. There is even an equation for this: the number of species in an area (S) is predicted from area (A) by $S = cA^z$, where c is a constant and z varies among species and habitats.

Overall, protecting core areas of high biodiversity, surrounded by buffer zones and linked by corridors, remains one of the highest priorities for the protection of wild species and ecological communities (Noss and Cooperrider 1994, Foreman 2004). Part of maintaining such wild places means allowing the natural filters, like fire and flooding, to continue operation to assemble these wild communities (Keddy et al. 2007, Keddy 2009). This is a topic to which we will return.

RESPONSE RULES AND INERTIA IN COMMUNITY ASSEMBLY (TIME DEPENDENCE)

In this book we have treated community assembly as an almost instantaneous process. We leap from the species pool and the trait matrix directly to the community vector using certain statistical tools. In practice, there are several limitations to this approach. First, as noted in the very first chapter, time seems to be an important part of community assembly, at least when we can observe the process of assembly actually occurring in communities. That is, few communities are suddenly assembled from the pool, but rather they arise slowly through immigration of propagules and various species interactions (a process sometimes known as succession). The concept of change through time, and ecological succession, has been an ongoing discussion in community ecology since Clements (1916), with themes including both the patterns that occur and the mechanisms that may be responsible, and even whether the concept should be discarded. Would it be beneficial to include time within community assembly? To what extent does time already appear, if we include dispersal-related traits in the trait matrix? These are two open questions. Returning to Figure 8.1, is it possible that assembly over time actually contains two quite different elements: the assembly of functional types and the accumulation of species within each functional type? And, if this is the case, might we consider each functional type in a community to operate independently, somewhat as an island, with species arriving and going locally extinct within each functional type? More open questions.

Time arises in another way. Consider the concept of response rules as opposed to assembly rules (Keddy 1992). The term response rules may be useful, and even necessary, since in most cases new communities are assembled from the remains of previous communities, somewhat similar in the way new cities are built upon the ruins of old cities. The list of exceptions to this generalization is rather short. Here are five: newly formed volcanic islands, debris fields from landslides, recently deposited sediments in flood plains, abandoned farm fields with a past history of herbicide use and recently abandoned mining sites. Such situations are relatively uncommon, overall. Even when forests are clear-cut, there are a variety of immediate sources of regeneration, including sprouting stumps (for certain species), buried rhizomes and buried seeds. Even after extreme natural events like hurricanes and ice storms, there are abundant sources of natural regeneration already present, again, in the case of plants, from root sprouts, buried rhizomes and seeds. Mycorrhizal networks are also already present. Both eggs and adults of many arthropods may survive these extreme events. In such cases, the new community vector is therefore assembled from two quite different pools: the pool of species already present and the larger pool of species which will disperse to the site. The pool of species already present at the site will have considerable advantages, including more or less immediate arrival (no dispersal delays), and often considerable kinds of pre-establishment in the form of root systems and storage structures such as rhizomes and tubers, as well as various linkages to mycorrhiza.

Therefore, in a majority of cases, assembly involves two sets of parts: those already present and the new ones that will arrive. To slightly switch analogies, from bookcases to houses, assembly rules might be compared to building a new house, while response rules might be compared to the process of renovations of an existing house. Having a set of interlinked parts from the previous community provides what we might wish to call *community inertia*. The same situation is true for much larger-scale issues, such as how communities will respond to climate change. A new forest, after all, will emerge not directly from the pool, but slowly change as the existing species decline in fitness relative to newly arrived species. These existing communities impose a considerable lag in response to changing filters. Moreover, the inertia, or lag, is mostly the result of a single plant organ – the meristem, which exists in about six styles of functional organs (Figure 8.7). Thus, the recent review of below-ground bud banks (Ott et al. 2019) might be a good starting point for studies of how to incorporate inertia into community assembly, and therefore into the development of response rules. In a satisfying way, Figure 8.7 takes us back in the history of ecology, and in the narrative structure of this book, to Raunkiaer and his work on the location of meristems (e.g., Figure 6.1) as a fundamental trait for community ecology. Even bird and moth communities will be affected by this simple trait in plants.

The topic of response rules would seem to be particularly important for predicting how communities will respond to widespread human effects upon landscapes. The obvious example is climate change, but overgrazing, deforestation and eutrophication are other widespread human effects upon landscapes, imposing anthropogenic filters that act upon existing communities and shift their composition in predictable directions. One could argue that in such cases, the process is better viewed as one of community response than

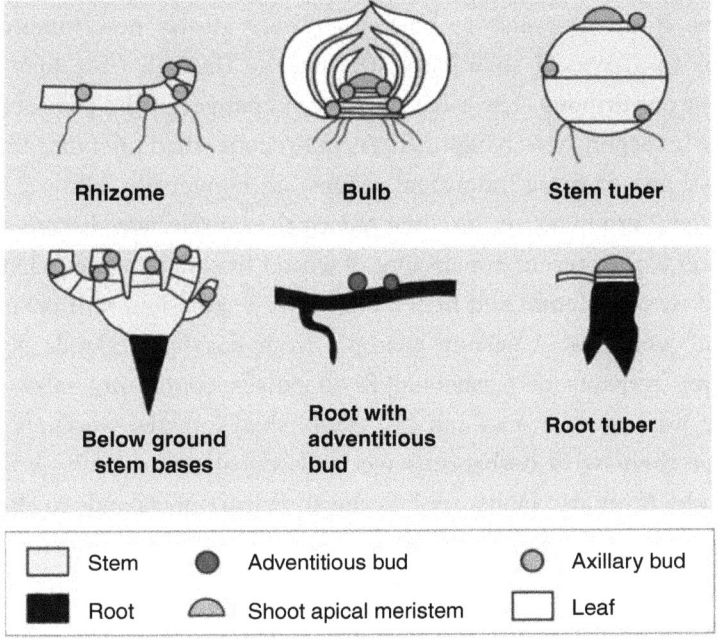

FIGURE 8.7 Not all ecological communities are instantaneously assembled from species pools. Six kinds of meristems are common in plant communities, and these produce inertia in response to changing environmental filters. (From Ott et al. 2019.)

community assembly. The filters responsible for these changes are often acting at large scales, regional and global. Hence, Paul has deliberately provided entire book chapters on the effects of filters, including nutrients and herbivores, on plant communities (Keddy 2017, chapters 3 and 6, respectively), consciously including a mix of older and newer studies that need consideration in future work. On the pressing topic of deforestation, Thirgood (1981) is still a classic book, extending the history of deforestation in European and African history back several millennia to Greek and Roman activities in the Mediterranean. Going back in time even further, human grazing during Neolithic times might have contributed to the formation of the modern Sahara Desert (Wright 2017).

Community responses may also be affected by functional groups, if the response to changing filters allows new functional groups to arrive. In such cases, the change through time may not even be continuous. The most significant changes in the community vector in response to changing filters may occur when tolerance limits remove one or more functional groups, or, conversely, allow a new functional group to invade. One practical example was discussed in Chapter 2: the presence or absence of winter frost determines whether or not woody plants, and hence mangrove vegetation, will occur in coastal wetlands. A second example from coastal wetlands is the extreme impacts of occasional flood pulses containing saltwater. These infrequent events can kill established forested wetlands and replace them with herbaceous wetlands (recall Figure 2.9). A third example, from arid landscapes, is the shift from grasslands to shrublands with overgrazing (Archer 1989, Wright 2017). Therefore, in general, filters that control the presence or absence of functional types, in this case woody plants, may be particularly important in community assembly.

To conclude this section, we suggest that the concept of inertia may be important in future studies of community assembly. Moreover, when we are dealing with filters that are greatly influenced by human populations, such as in climate change, overgrazing and eutrophication, we may need to include change with time and inertia in models that predict how communities respond to changing filters. You may wish to consider how you would take the CATS or Traitspace models, or even an entirely new model, to pose the following questions. How would the vegetation of northern Africa change if grazing pressure were reduced to sustainable levels? How would the vegetation of the Rocky Mountains change if global temperatures increase? And how will the communities of insects, reptiles, birds, mammals and fungi re-form in response to such changes?

THE ROLE OF CAUSAL FACTORS IN ECOLOGICAL COMMUNITIES AND RESTORATION

Now for the last section. We have seen how filters drive the assembly of communities, overall. And we have just observed that humans are changing filters at global scales by altering the climate, increasing grazing pressures and elevating nutrient levels in communities. This is not good news. It is why some scientists propose naming a new geological era in honour of our own species: the Anthropocene. At the same time, there is some good news: humans are also learning how to change filters in order to restore ecological communities to more desirable compositions (Cairns 1980, Allison and Murphy 2017). In either case, degradation or restoration, it would seem that filters deserve proper and rigorous attention. Indeed, the entire field of degradation and restoration of ecosystems would seem to be based upon the manipulation of filters (Figure 8.8).

Probably the most important point, and one we have already made in Chapter 2, deserves repeating. In most habitats, a relatively small number of filters is likely responsible for creating most communities. Alas, given the nature of ecological communities, and the habitual patterns of scientists, it is all too easy to go into the field and ask "Does filter X affect community Y?" It is in the very nature of ecological systems that if we select almost any filter and modify it enough, we are likely to find an effect of some sort upon some species. Therefore, asking this question seems once again to take us to Zeno's

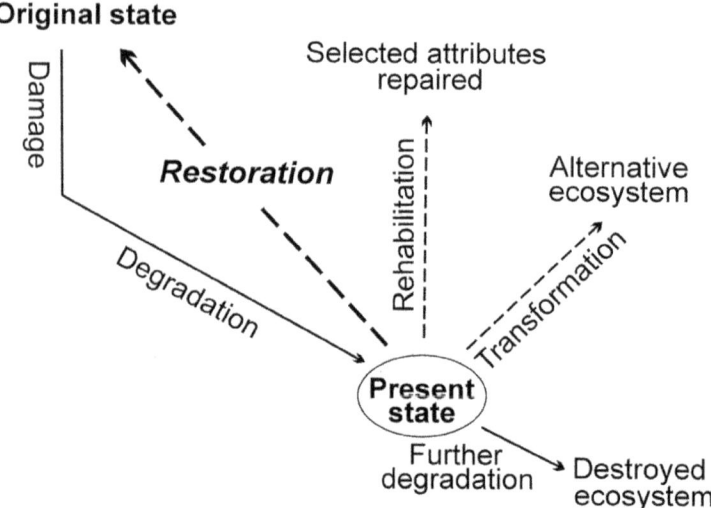

FIGURE 8.8 Both the degradation and the restoration of ecological communities is driven by changes in ecological filters. Examples include changing climate, overgrazing by goats, eutrophication and changing fire and flood regimes. A few filters likely dominate each community. (From Keddy 2017, figure 13.16, after Cairns 1980).

now well-worn tortoise. It seems a different question is needed: "Which environmental filters are most important in creating a particular community?" A subtle change in emphasis of wording, and much more work, perhaps, but the answer is far more useful. In wetlands, for example (Table 2.3), hydrology and fertility are likely the predominant filters, although, as noted, the list can be extended to include fire, competition, grazing, burial, roads and more. To develop good general models for predicting community assembly, we need to rank the filters in order of importance, and put the first ones first in our work. Hence, in this book, we have put a good deal of emphasis upon drought and flooding. Once ecologists have identified the most important filters, we can then explore how these essential causal factors are leading to degradation of communities, and consider how to modify those same filters in order to restore those same communities. Indeed, this topic is so important that we offered two propositions in Chapter 2 on this very point: (1) only a small number of filters is likely important in a particular community and (2) a large number of species are likely controlled by the same filters (Table 8.1). In Chapter 2 we have already written about the importance of factorial design experiments for measuring the relative importance of filters, and actually concluded that chapter with their application to the restoration of coastal wetlands.

We have placed a good deal of emphasis upon physical filters, including drought, flooding and fire. Yet, biological interactions are also important: predation, grazing and competition all have the potential to act as filters. In this regard, we remind younger ecologists that many of the world's ecological communities have been significantly disrupted by altering predation and herbivory. Top predators are being removed from communities and landscapes around the world (Ripple et al. 2014). A top priority for restoration is to return the top predators to ecosystems, that is, to make more space for populations of alligators, wolves, coyotes, lions, tigers and sharks.

The case with herbivores is slightly more complicated. Some communities and landscapes are being disrupted by abnormally high populations of herbivores: the world has too many goats, sheep, cattle and deer. Other communities are being disrupted by the loss of native herbivores, including the historical slaughter of bison in North America and the ongoing slaughter of elephants in Africa.

Competition can also act as a biological filter, particularly in communities of sessile organisms where space is limited. A great deal has been written about the role of competition in ecological communities. Robert MacArthur (1972) thought that biogeographic patterns were a good source of evidence for competition, leading to a great many studies of pattern that presumed competition as the causal mechanism without actually demonstrating it (Weiher and Keddy 1999b). To repeat, just because there is a pattern in nature, be it in plants in quadrats or birds in archipelagoes, does not mean it was caused by competition. Daniel Simberloff (1984) once called competition "the great god" in community ecology, presumably in reference to its presumed power combined with the ambiguity of evidence for its presence. Yes, in some cases competition is a powerful filter, and there are good grounds to conclude that it plays an important role in assembling communities, at some times and in some locations. At the same time, the evidence suggests that the importance of competition varies with time, space and scale, and sometimes it may not be among the most important filters. We have noted with some surprise that the CATS and Traitspace models work relatively well in predicting ecological communities in spite of the fact that they entirely ignore the presumed effects of competition. Perhaps some clarity will come when we begin to view competition as one factor among many, and when we attempt to measure its importance relative to other filters. Thus, there is an emerging literature on the importance of competition, dealing with fundamental issues such as what we mean by importance and how we might measure it (Weldon and Slauson 1986, Keddy 2001, Diaz-Sierra et al. 2016).

There are some obvious immediate problems that we can immediately address. We can (1) allow increased populations of top-level predators, and (2) reduce the populations of herbivores causing anthropogenic overgrazing. Oddly enough, the origins of both of these problems likely arise out of human behaviour: anyone who has worked with these two filters will have personal experience with human fear of predators, and human desire for large herds of herbivores. Regarding the former, many more people are killed by stepladders and power tools than by alligators or wolves, but the deaths of people falling off stepladders is mostly ignored while the death of one person by an alligator makes front-page news. This illustrates the context in which we often find ourselves when working in restoration ecology: the study of ecological assembly and community restoration is enmeshed in the context of human psychology. So, we repeatedly encounter human psychological traits including folly (Tuchman 1984), wilful blindness (Heffernan 2011) and craving (Keddy 2020), and it is these behaviours that often drive the dysfunction we see in landscapes and communities. So, when we are trying to restore certain ecological filters, we often find that the problems we have to address deal mostly with human psychological baggage that likely goes back to Paleolithic societies. For example, Figure 2.9 illustrates the primary filters and plant community responses to restoration of coastal Louisiana, but the current dysfunctional state of the coastline is being driven by psychological factors, including folly, wilful blindness and greed. So, it is important to tease apart the ecological filters and the human behaviours that create them. In Chapter 2 we wrote about the importance of grazing as an environmental filter, and have suggested elsewhere that overgrazing by herds of goats is a particular threat to arid lands around the world. When we apply the CATS model to Morocco, it describes an area already very much changed by grazing. Louisiana and Morocco are therefore vivid examples that ecologists are testing their models in landscapes in which filters have been significantly changed in recent history. In the same way, most modern ecologists are similarly working in landscapes from which predators have been removed, possibly leading to highly disrupted food webs overall.

In this case, islands have something useful to say to the field of ecology as a whole. We now have good evidence about just how damaging herbivores, particularly introduced goats, can be to island vegetation, and an increasing number of satisfying examples of how positive changes occur in the communities and landscapes once goats are removed (Hamann 1993, Cruz et al. 2009).

The field of restoration is experiencing an existential crisis of sorts, again owing to human behaviour. The classic model of restoration, as illustrated in Figure 8.8, proposes that the goal of restoration is to return an ecosystem to a pre-disturbed historical state. The search for reference conditions that describe these pre-disturbance states was a fundamental step in ecological restoration. But reference conditions may be difficult or impossible to define in regions that have been degraded for multiple centuries (e.g., Europe), and climate change may have moved us into an entirely new climatic state (Harris et al.

2006, Laughlin et al. 2017). One could argue that restoring historical community composition would set the restoration project up for failure if the species that are selected for planting are not adapted to the new climatic conditions.

The search for reference conditions was defined by asking the question, "What species were present historically?" Daniel suggests that is not quite the right question to ask in a time of rapid environmental change. Rather, we should be asking "*Why* were those species present historically?" The answer, presumably, is that species were present historically because they exhibited traits that were adapted to those conditions. Therefore, if we understand the filters (Chapter 2) and how they shape the relationships between traits and environments (Chapter 5), then the task of restoration is transformed. The task is no longer determining the list of species that were present historically, the task becomes determining the traits that are good matches to the new conditions. Of course, an understanding of past conditions allows us to put the proposed new conditions into context: perhaps analogous communities existed at other times or other places. For example, although it may be difficult to imagine, not that long ago, between 5,000 and 15,000 years BP, the enormous Sahara Desert was nearly completely vegetated with annual grasses and shrubs (deMenocal et al. 2000), as well as *Acacia* trees and scattered populations of tropical trees (Kröpelin et al. 2008). There were numerous lakes and large animals, including antelope, giraffes, elephants, hippopotamuses and crocodiles. Moreover, the change from this humid state to the current arid state occurred abruptly, over decades to centuries, when summer insolation passed a threshold value. These changes in climate also changed lake salinity, causing abrupt changes in the aquatic fauna (Kröpelin et al. 2008). It is therefore possible to consider northern Africa to have a bistable state, with "rapid and large-amplitude climate transitions" (deMenocal et al. 2000). Although the changes are thought to be driven mostly by orbital forcing (changes in the seasonal distribution of solar radiation), there are biological feedbacks driven by changes in albedo from vegetation.

Of course, we can also be cautious about invoking a single cause to explain current changes in community composition. It is important to entertain multiple working hypotheses in general, and to be cognizant that while the list of important filters may be short, several can usually be identified, and they are operating simultaneously (Chapter 2). We need to be careful not to fall into either/or thinking. In the case of North Africa, we should note that anthropogenic factors are also likely to be driving recent changes in community composition. These anthropogenic factors include overpopulation, deforestation, overgrazing and ancient warfare (Thirgood 1981, Keddy 2017, pp. 492–494, 506–507), as well as modern warfare. Moreover, Wright (2017) presents evidence that pre-historic grazing practices may have contributed to the loss of vegetation in the Sahara at the end of the last humid period.

Given the changes in global climate, and the fact that they may be driven by both astronomical and anthropogenic factors, restoration ecologists may have to place more emphasis upon determining which traits will increase the likelihood of achieving desirable outcomes. For example, if the outcome is to restore pollinator habitat, then floral traits can be used to derive an assemblage of species to achieve that objective. At a more local scale, many of the local restoration projects cite the desire to increase the abundance of pollinators, such as butterflies and birds, because they bring joy to local communities of people. If the outcome is to achieve a resilient and productive community in the face of impending drought, then drought tolerance traits such as xylem embolism resistance or leaf turgor loss point can be used to derive an assemblage of species to achieve that objective. In fact, the system of linear equations used in the CATS model can be modified to produce such assemblages for ecological restoration (Laughlin et al. 2018a), and, as we write this, Daniel and his colleagues are testing these new ideas experimentally in the grasslands of Wyoming and California.

Thus, ecological restoration can be viewed as not a separate applied field, but simply as the enlightened application of community assembly rules to re-establish desirable ecological communities. The primary tool of restoration is manipulation of filters that are causing dysfunction, and replacing them with filters that lead to recovery (Figures 2.9 and 8.8). This process may also involve augmenting dispersal of species that are better adapted to the new conditions by supplying propagules, or reintroducing adults (e.g., planting container-grown plants, releasing captive-bred animals). Although there are now entire books and many training programs for ecological restoration, we would like to observe here that the most important step is *restoring the natural filters* that historically shaped the original ecosystems. This often requires experiments to first determine which species may be better adapted to the new changed conditions. Once these filters are restored, the community should naturally shift in the desired direction and the new species will maintain viable populations. Thus, we close with the observation that community ecology, as we have described it in this book, is really the underlying theoretical framework for restoration ecology (Temperton et al. 2004, Funk et al. 2008). And, when properly carried out, ecological restoration will allow further advances in community ecology by providing concrete information on the relative importance of environmental filters, and the relationships between traits and those filters.

We have come a long way from Chapter 1. We have introduced some foundational figures in the field, including Raunkiaer, Tansley and Rigler. We have reviewed the key elements of community ecology: filters (Chapter 2), species pools (Chapter 3) and traits (Chapter 4). We explored ways to quantify trait–environment relationships (Chapter 5) and considered how functional groups fit into the story (Chapter 6). We have introduced the two models currently used to assemble communities from pools using traits (Chapter 7). We have offered six propositions to guide future development of ecological assembly rules (Chapters 2, 5, and 8, Table 8.1). And, we have allowed ourselves this final chapter to talk more speculatively about four areas that we think deserve further emphasis in future work: assembly of functional groups, rarity, response rules, and restoration. It is now time to bring this project to a close, let Zeno's poor tortoise have a well-earned rest in the shade, and invite younger scholars to move this project forward. We look forward to better predictive models and more wild landscapes.

KEY POINTS OF THE CHAPTER

- Community ecology has a logical structure based upon species pools, traits and filters.
- Proposition 5: A community is *functionally assembled* when there is at least one species representing each functional type that is adapted to the habitat.
- Proposition 6: A community is *fully assembled* when each functional type has the maximum number of species that can coexist.
- Functional groups may have an important place in the assembly of communities and predicting abundances of functional groups may be easier than for species. We use an example from an ephemeral wetland to demonstrate this point.
- A large proportion of biological diversity lies in the lower tail of the log-normal distribution, yet ecologists frequently trim this tail from their data sets in order to construct models.
- The rising tide of species extinctions requires us to consider how to include the conservation of rare species in models for community assembly.
- Most communities arise out of pre-existing communities, so when filters change there is often inertia in community response.
- Community assembly rules lie at the heart of ecological restoration.

References

Ackerly, D. D., and W. K. Cornwell. 2007. A trait-based approach to community assembly: partitioning of species trait values into within- and among-community components. *Ecology Letters* **10**: 135–145.

Adler, P. B., D. Smull, K. H. Beard, et al. 2018. Competition and coexistence in plant communities: intraspecific competition is stronger than interspecific competition. *Ecology Letters* **21**: 1319–1329.

Allen, M. F., W. Swenson, J. I. Querejeta, L. M. Egerton-Warburton and K. K. Treseder. 2003. Ecology of mycorrhizae: a conceptual framework for complex interactions among plants and fungi. *Annual Review of Phytopathology* **41**: 271–303.

Allison, S. K., and S. D. Murphy (eds.). 2017. *Routledge Handbook of Ecological and Environmental Restoration*. Routledge, New York.

Anderson, M. J., T. O. Crist, J. M. Chase, et al. 2011. Navigating the multiple meanings of β diversity: a roadmap for the practicing ecologist. *Ecology Letters* **14**: 19–28.

Anjos, M. B., R. R. De Oliveira and J. Zuanon. 2008. Hypoxic environments as refuge against predatory fish in the Amazonian floodplains. *Brazilian Journal of Biology* **68**: 45–50.

Archer, S. 1989. Have southern Texas savannas been converted to woodlands in recent history? *The American Naturalist* **134**: 545–561.

Archibold, O. W. 1995. *Ecology of World Vegetation*. Chapman and Hall, London.

Arendt, W. J. 1988. Range expansion of the Cattle Egret (*Bubulcus ibis*) in the Greater Caribbean Basin. *Colonial Waterbirds* **11**: 252–262.

Armstrong, W., and J. Armstrong. 2005. Stem photosynthesis not pressurized ventilation is responsible for light-enhanced oxygen supply to submerged roots of alder (*Alnus glutinosa*). *Annals of Botany* **96**: 591–612.

Axelrod, D. I. 1985. Rise of the grassland biome, central North America. *The Botanical Review* **51**: 163–201.

Badiou, P., L. G. Goldsborough and D. Wrubleski. 2011. Impacts of Common Carp (*Cyprinus carpio*) on freshwater ecosystems: a review. Pages 121–146 in *Carp: Habitat, Management and Diseases*. J. D. Sanders and S. B. Peterson (eds.). Nova Science Publishers, New York.

Bartlett, M. K., C. Scoffoni, R. Ardy, et al. 2012. Rapid determination of comparative drought tolerance traits: using an osmometer to predict turgor loss point. *Methods in Ecology and Evolution* **3**: 880–888.

Bartram, W. 1791. *Travels Through North and South Carolina, Georgia, East and West Florida, the Cherokee Country, the Extensive Territories of the Muscogulges, or Creek Confederacy, and the Country of the Chactaws; Containing an Account of the Soil and Natural Productions of Those Regions, Together with Observations on the Manners of the Indians.* James and Johnson, Philadelphia, PA. (Electronic version, 2001, Documenting the American South, University of North Carolina at Chapel Hill.)

Baskin C. C., and J. M. Baskin. 1998. *Seeds: Ecology, Biogeography, and Evolution of Dormancy and Germination.* Academic Press, San Diego, CA.

Baskin, C., and J. Baskin. 2014. *Seeds: Ecology, Biogeogaphy, and Evolution of Dormancy and Germination*, 2nd ed. Academic Press, San Diego, CA.

Beech, E., M. Rivers, S. Oldfield and P. P. Smith. 2017. GlobalTreeSearch: the first complete global database of tree species and country distributions. *Journal of Sustainable Forestry* **36**: 454–459.

de Bello, F., W. Thuiller, J. Lepš, et al. 2009. Partitioning of functional diversity reveals the scale and extent of trait convergence and divergence. *Journal of Vegetation Science* **20**: 475–486.

de Bello, F., S. Lavorel, S. Lavergne, et al. 2013. Hierarchical effects of environmental filters on the functional structure of plant communities: a case study in the French Alps. *Ecography* **36**: 393–402.

Belsky, A. J. 1992. Effects of grazing, competition, disturbance and fire on species composition and diversity in grassland communities. *Journal of Vegetation Science* **3**: 187–200.

Bergmann, J., A. Weigelt, F. van der Plas, et al. 2020. The fungal collaboration gradient dominates the root economics space in plants. *Science Advances* **6**: eaba3756.

Bertness, M. D. 1991. Interspecific interactions among high marsh perennials in a New England Salt marsh. *Ecology* **72**: 125–137.

Biswell, H. H. 1974. Effect of fire on chaparral. Pages 321–364 in *Fire and Ecosystems*. T. T. Kozlowski and C. E. Ahlgren (eds.). Academic Press, New York.

Blaney, C. S., and D. M. Mazerolle. 2009. *Rare plant inventory of lakes in the Ponhook – Molega Lakes region. Report to the Endangered Species Recovery Fund and Nova Scotia Species at Risk Conservation Fund.* Atlantic Canada Conservation Data Centre, Sackville, NB.

Bleakney, J. S. 1958. *A Zoogeographical Study of the Amphibians and Reptiles of Eastern Canada*. National Museum of Canada Bulletin No. 155. Queen's Printer, Ottawa.

Blonder, B., R. E. Kapas, R. M. Dalton, et al. 2018. Microenvironment and functional-trait context dependence predict alpine plant community dynamics. *Journal of Ecology* **106**: 1323–1337.

Blossey, B., M. Schwarzländer, P. Häfliger, R. Casagrande and L. Tewksbury. 2002. Common reed. Pages 131–138. in *Biological Control of Invasive Plants in the Eastern United States*. F. V. Driesche, B. Blossey, M. Hoodle, S. Lyon and R. Reardon (eds.). United States Department of Agriculture Forest Service, Morgantown, WV.

Boesch, D. F., M. N. Josselyn, A. J. Mehta, et al. 1994. Scientific assessment of coastal wetland loss, restoration and management in Louisiana. *Journal of Coastal Research*, Special Issue 20.

Bond, W. J., and J. E. Keeley. 2005. Fire as a global "herbivore": the ecology and evolution of flammable ecosystems. *Trends in Ecology and Evolution* **20**: 387–394.

Bond, W. J., F. I. Woodward and G. F. Midgley. 2005. The global distribution of ecosystems in a world without fire. *New Phytologist* **165**: 525–538.

Boonzaier, L., and D. Pauly. 2016. Marine protection targets: an updated assessment of global progress. *Oryx* **50**: 27–35.

Boutin, C., and P. A. Keddy. 1993. A functional classification of wetland plants. *Journal of Vegetation Science* **4**: 591–600.

Bowles, M., M. Jones, L. Wetstein, R. Hyerczk and K. Klick. 1994. *Results of a Systematic Search for* Thismia americana Pfeiffer *in Illinois*. Morton Arboretum, Lisle, IL.

Bradley, D. 2013. *Southern Ontario Vascular Plant Species List*, 3rd ed. Southern Science and Information Section Ontario Ministry of Natural Resources, Peterborough, ON.

Brejão, G. L., P. Gerhard and J. Zuanon. 2013. Functional trophic composition of the ichthyofauna of forest streams in eastern Brazilian Amazon. *Neotropical Ichthyology* **11**: 361–373.

Brodribb, T. J., J. Pittermann and D. A. Coomes. 2012. Elegance versus speed: examining the competition between conifer and angiosperm trees. *International Journal of Plant Sciences* **173**: 673–694.

Brown, A. M., D. I. Warton, N. R. Andrew, et al. 2014. The fourth-corner solution: using predictive models to understand how species traits interact with the environment. *Methods in Ecology and Evolution* **5**: 344–352.

Brundrett, M. C. 2009. Mycorrhizal associations and other means of nutrition of vascular plants: understanding the global diversity of host plants by resolving conflicting information and developing reliable means of diagnosis. *Plant and Soil* **320**: 37–77.

Cadotte, M. W., and T. J. Davies. 2016. *Phylogenies in Ecology: A Guide to Concepts and Methods*. Princeton University Press, Princeton, NJ.

Cairns, J. (ed.) 1980. *The Recovery Process in Damaged Ecosystems*. Ann Arbor Science, Ann Arbor, MI.

Caliński, T., and J. Harabasz. 1974. A dendrite method for cluster analysis. *Communications in Statistics: Theory and Methods* **3**: 1–27.

Cameron, A. W. 1958. *Mammals of the Islands in the. Gulf of St. Lawrence*. Canada Department of Northern Affairs and National Resources, Ottawa.

Carmona, C. P., R. Tamme, M. Pärtel, et al. 2021. Erosion of global functional diversity across the tree of life. *Science Advances* **7**: eabf2675.

Carrascal, L. M., E. Moreno and J. L. Telleria. 1990. Ecomorphological relationships in a group of insectivorous birds of temperate forests in winter. *Holarctic Ecology* **13**: 105–111.

Carson, W. P., and S. T. A. Pickett. 1990. Role of resources and disturbance in the organization of an old-field plant community. *Ecology* **71**: 226–238.

Casatti, L., and R. M. C. Castro. 2006. Testing the ecomorphological hypothesis in a headwater riffles fish assemblage of the rio São Francisco, southeastern Brazil. *Neotropical Ichthyology* **4**: 203–214.

Catling, P. M., and V. R. Brownell. 1999a. The flora and ecology of southern Ontario granite barrens. Pages 392–405 in *Savannas, Barrens, and Rock Outcrop Plant Communities of North America*. Anderson, R. C., J. S. Fralish and J. M. Baskin (eds). Cambridge University Press, Cambridge.

Catling, P. M., and V. R. Brownell. 1999b. Alvars of the Great Lakes region. Pages 375–391 in *Savannas, Barrens, and Rock Outcrop Plant Communities of North America*. Anderson, R. C., J. S. Fralish and J. M. Baskin (eds). Cambridge University Press, Cambridge.

Catling, P. M., and W. G. Dore. 1982. Status and identification of *Hydrocharis morus-ranae* and *Limnobium spongia* (Hydrocharitaceae) in northeastern North America. *Rhodora* **84**: 523–545.

Catling, P. M., J. E. Cruise, K. L. McIntosh and S. M. McKay. 1975. Alvar vegetation in Southern Ontario. *Ontario Field Biologist* **29**(2): 1–25.

Catling, P. M., K. W. Spicer, and L. P. Lefkovitch, 1988. Effects of the introduced floating vascular aquatic, *Hydrocharis morsus-ranae* (Hydrocharitaceae), on some North American aquatic macrophytes. *Naturaliste Canadien* **115**: 131–137.

Chaneton, E. J., and J. M. Facelli. 1991. Disturbance effects on plant community diversity: spatial scales and dominance hierarchies. *Vegetatio* **93**: 143–156.

Chapin, F. S. 1980. The mineral nutrition of wild plants. *Annual Review of Ecology and Systematics* **11**: 233–260.

Cherlet, M., C. Hutchinson, J. Reynolds, et al. (eds.). 2018. *World Atlas of Desertification*. Publication Office of the European Union, Luxembourg.

Chirima, G. J., N. Owen-Smith and B. F. N. Erasmus. 2012. Changing distributions of larger ungulates in the Kruger National Park from ecological aerial survey data. *Koedoe* **54**: 24–35.

Choat, B., S. Jansen, T. J. Brodribb, et al. 2012. Global convergence in the vulnerability of forests to drought. *Nature* **491**: 752–755.

Christian, B., and T. Griffiths. 2016. *Algorithms to Live By: The Computer Science of Human Decisions*. Allen Lane, Penguin Canada, Toronto.

Clark, C. W. 1973. The economics of overexploitation. *Science* **181**: 630–634.

Clark, M. A., J. Siegrist and P. A. Keddy. 2008. Patterns of frequency in species-rich vegetation in pine savannas: effects of soil moisture and scale. *Ecoscience* **15**: 529–535.

Clatworthy, J. N., and J. L. Harper. 1962. The comparative biology of closely related species living in the same area: V. Inter- and intraspecific interference within cultures of *Lemna* spp. and *Salvinia natans*. *Journal of Experimental Botany* **13**: 307–324.

Clements, F. E. 1916. *Plant Succession: An Analysis of the Development of Vegetation*. Carnegie Institution of Washington, DC.

Clements, F. E. 1933. *Competition in Plant Societies*. Carnegie Institution of Washington, Washington, DC.

Clements, F. E., J. E. Weaver and H. C. Hanson. 1929. *Plant Competition*. Carnegie Institution of Washington, Washington, DC.

Cole, L. C. 1949. The measurement of interspecific association. *Ecology* **30**: 411–424.

Coley, P. D. 1983. Herbivory and defensive characteristics of tree species in a lowland tropical forest. *Ecological Monographs* **53**: 209–233.

Colinvaux, P. 1978. *Why Big Fierce Animals Are Rare: An Ecologist's Perspective*. Princeton University Press, Princeton, NJ.

Colinvaux, P. 1986. *Ecology*. Wiley and Sons, Toronto.

Colwell, R. K., and J. A. Coddington. 1994. Estimating terrestrial biodiversity through extrapolation. *Philosophical Transactions of the Royal Society of London, Series B* **345**: 101–118.

Connell, J. H. 1978. Diversity in tropical rain forests and coral reefs. *Science* **199**: 1302–1310.

Cornell, H. V., and S. P. Harrison. 2014. What are species pools and when are they important? *Annual Review of Ecology, Evolution and Systematics* **45**: 45–67.

Cornwell, W. K., and D. D. Ackerly. 2009. Community assembly and shifts in plant trait distributions across an environmental gradient in coastal California. *Ecological Monographs* **79**: 109–126.

Cornwell, W. K., and D. D. Ackerly. 2010. A link between plant traits and abundance: evidence from coastal California woody plants. *Journal of Ecology* **98**: 814–821.

COSEWIC 2010. *COSEWIC Assessment and Status Report on the Eastern Mountain Avens* Geum peckii *in Canada.* Committee on the Status of Endangered Wildlife in Canada, Ottawa.

Cowling, R. M., P. W. Rundel, B. B. Lamont, M. K. Arroyo and M. Arianoutsou. 1996. Plant diversity in Mediterranean-climate regions. *Trends in Ecology and Evolution* **11**: 362–366.

Crawford, R. M. M. 1982. Physiological response to flooding. Pages. 453–77 in *Encyclopedia of Plant Physiology.* O. L. Large, P. S. Nobel, C. B. Osmond and H. Ziegler (eds.). Springer-Verlag, Berlin.

Crins, W. J., P. A. Gray, P. W. C. Uhlig and M. C. Wester. 2009. *The Ecosystems of Ontario, Part 1: Ecozones and Ecoregions.* Ontario Ministry of Natural Resources, Peterborough, ON.

Cruz, F., V. Carrion, K. J. Campbell, C. Lavoie and C. J. Donlan. 2009. Bio-economics of large-scale eradication of feral goats from Santiago Island, Galápagos. *Journal of Wildlife Management* **73**: 191–200.

Cummins, K. W. 1973. Trophic relationships of aquatic insects. *Annual Review of Entomology* **18**: 83–206.

Cummins, K. W., and M. J. Klug. 1979. Feeding ecology of stream invertebrates. *Annual Review of Ecology and Systematics* **10**: 147–172.

Cummins, K. W., R. W. Merritt and P. C. N. Andrade. 2005. The use of invertebrate functional groups to characterize ecosystem attributes in selected streams and rivers in south Brazil. *Studies on Neotropical Fauna and Environment* **40**: 69–89.

Dale, M. 1999. *Spatial Pattern Analysis in Plant Ecology.* Cambridge University Press, Cambridge.

Dansereau, P. 1959. Vascular aquatic plant communities of southern Quebec: a preliminary analysis. *Transactions of the Northeast Wildlife Conference* **10**: 27–54.

Darwin, C. R. 1855. Effect of salt-water on the germination of seeds. *Gardeners' Chronicle and Agricultural Gazette* **47** (24 November): 773. From *The Complete Works of Darwin Online* (http://darwin-online.org.uk).

Daubenmire, R. F. 1968. The ecology of fire in grasslands. *Advances in Ecological Research* **5**: 209–266.

Davidson, D. W., R. S. Inouye and J. H. Brown 1984. Granivory in a desert ecosystem: experimental evidence for indirect facilitation of ants by rodents. *Ecology* **65**: 1780–1786.

Dawkins, R. 1996. *Climbing Mount Improbable*. W.W. Norton, New York.

Day, R. T., P. A. Keddy, J. McNeill and T. Carleton. 1988. Fertility and disturbance gradients: a summary model for riverine marsh vegetation. *Ecology* **69**: 1044–54.

deMenocal, P., J. Ortiz, T. Guilderson, et al. 2000. Abrupt onset and termination of the African Humid Period: rapid climate responses to gradual insolation forcing. *Quaternary Science Reviews* **19**: 347–361.

Diamond, J. M. 1975. Assembly of species communities. Pages 342–444 in *Ecology and Evolution of Communities*. M. L. Cody and J. M. Diamond (eds.). Belknap Press of Harvard University Press, Cambridge, MA.

Diamond, J., and T. J. Case (eds.). 1986. *Community Ecology*. Harper and Row, New York.

Díaz, S., J. Kattge, J. H. C. Cornelissen, et al. 2016. The global spectrum of plant form and function. *Nature* **529**: 167–171.

Diaz-Sierra, R., M. Verwijmeren, M. Rietkerk, V. Resco de Dios and M. Baudena. 2016. A new family of standardized and symmetric indices for measuring the intensity and importance of plant neighbour effects. *Methods in Ecology and Evolution* **8**: 580–591.

Digby, P. G. N., and R. A. Kempton. 1987. *Multivariate Analysis of Ecological Communities*. Chapman and Hall, London.

Dlott, F., and R. Turkington. 2000. Regulation of boreal forest understory vegetation: the roles of resources and herbivores. *Plant Ecology* **151**: 239–251.

Dray, S., and P. Legendre. 2008. Testing the species traits–environment relationships: the fourth-corner problem revisited. *Ecology* **89**: 3400–3412.

Duchesne, L. C., and D. W. Larson. 1989. Cellulose and the evolution of plant life. *Bioscience* **39**: 238–241.

Edmonds. J. (ed.) 1997. *Oxford Atlas of Exploration*. Oxford University Press, New York.

Ehrlich, A., and P. Ehrlich. 1981. *Extinction: The Causes and Consequences of the Disappearance of Species*. Random House, New York.

Ejrnæs, R., H. H. Bruun and B. J. Graae. 2006. Community assembly in experimental grasslands: suitable environment or timely arrival? *Ecology* **87**: 1225–1233.

Ellenberg, H., H. E. Weber, R. Düll, et al. 1991. Zeigwerte von Pflanzen in MittelEuropa. *Scripta Geobotanica* **18**: 1–248.

Ellner, S. P., D. Z. Childs and M. Rees. 2016. *Data-Driven Modelling of Structured Populations: A Practical Guide to the Integral Projection Model*. Springer, Gland.

Enquist, B. J., J. Norberg, S. P. Bonser, et al. 2015. Scaling from traits to ecosystems: developing a general trait driver theory via integrating trait-based and metabolic scaling theories. *Advances in Ecological Research* **52**: 249–318.

Eriksson, O. 1993. The species-pool hypothesis and plant community diversity. *Oikos* **68**: 371–374.

Fan, K., J. Tao, L. Zang, et al. 2019. Changes in plant functional groups during secondary succession in a tropical montane rain forest. *Forests* **10**. doi:10.3390/f10121134.

Fernald, M. L. 1921. The Gray Herbarium expedition to Nova Scotia 1920. *Rhodora* **23**: 89–111, 130–171, 184–195, 233–245, 257–278, 284–300.

Fierer, N., M. S. Strickland, D. Liptzin, M. A. Bradford and C. C. Cleveland. 2009. Global patterns in belowground communities. *Ecology Letters* **12**: 1238–1249.

Foreman, D. 2004. *Rewilding North America : A Vision for Conservation in the 21st Century*. Island Press, Washington, DC.

Foster, A. S., and E. M. Gifford, Jr. 1974. *Comparative Morphology of Vascular Plants*, 2nd ed. W. H. Freeman, San Francisco, CA.

Fowells, H. A. 1965. *Silvics of Forest Trees of the United States*. United States Department of Agriculture, Forest Service, Washington, DC.

Franco, A. C., and P. S. Nobel. 1989. Effect of nurse plants on the microhabit and growth of cacti. *Journal of Ecology* **77**: 870–886.

Fraser, L. H., J. Pither, A. Jentsch, et al. 2015. Worldwide evidence of a unimodal relationship between productivity and plant species richness. *Science* **349**: 302–305.

Frenette-Dussault, C., B. Shipley, D. Meziane and Y. Hingrat. 2013. Trait-based climate change predictions of plant community structure in arid steppes. *Journal of Ecology* **101**: 484–492.

Frid, C. J. L. 1989. The role of recolonization processes in benthic communities, with special reference to the interpretation of predator-induced effects. *Journal of Experimental Marine Biology and Ecology* **126**: 163–171.

Funk, J. L., E. E. Cleland, K. N. Suding and E. S. Zavaleta. 2008. Restoration through reassembly: plant traits and invasion resistance. *Trends in Ecology and Evolution* **23**: 695–703.

Funk, J. L., J. E. Larson, G. M. Ames, et al. 2017. Revisiting the Holy Grail: using plant functional traits to understand ecological processes. *Biological Reviews* **92**: 1156–1173.

Fyllas, N. M., C. A. Quesada and J. Lloyd. 2012. Deriving plant functional types for Amazonian forests for use in vegetation dynamics models. *Perspectives in Plant Ecology, Evolution and Systematics* **14**: 97–110.

Garnier, E., M.-L. Navas and K. Grigulis. 2016. *Plant Functional Diversity: Organism Traits, Community Structure, and Ecosystem Properties*. Oxford University Press, Oxford.

Gauch, H. G., Jr. 1982. *Multivariate Analysis in Community Ecology*. Cambridge University Press, Cambridge.

Gaudet, C. L., and P. A. Keddy. 1988. Predicting competitive ability from plant traits: a comparative approach. *Nature* **334**: 242–243.

Geho, E. M., D. Campbell and P. A. Keddy. 2007. Quantifying ecological filters: the relative impact of herbivory, neighbours, and sediment on an oligohaline marsh. *Oikos* **116**: 1006–1016.

Gelman, A., and J. Hill. 2006. *Data Analysis Using Regression and Multilevel/ Hierarchical Models*. Cambridge University Press, New York.

Gertenbach, W. P. D. 1983. Landscapes of the Kruger National Park. *Koedoe* **26**: 9–121.

Gibson, A. C., and P. S. Nobel. 1986. *The Cactus Primer*. Harvard University Press, Cambridge, MA.

Gilpin, M. E., M. P. Carpenter and M. J. Pomerantz. 1986. The assembly of a laboratory community: multispecies competition in *Drosophila*. Pages 23–40 in *Community Ecology*. J. Diamond and T. J. Case (eds.). Harper and Row, New York.

Glitzenstein, J. S., D. R. Streng and D. D. Wade. 2003. Fire frequency effects on longleaf pine (*Pinus palustris* P. Miller) vegetation in South Carolina and northeast Florida, USA. *Natural Areas Journal* **23**: 22–37.

Gnanadesikan, R. 1997. *Methods for Statistical Data Analysis of Multivariate Observations*, 2nd ed. Wiley, New York.

Goldsmith, F. B., and C. M. Harrison. 1976. *Description and analysis of vegetation*. Pages 85–155 in *Methods in Plant Ecology*. S.B. Chapman (ed.). Blackwell Scientific, Oxford.

Götzenberger, L., F. de Bello, K. A. Bråthen, et al. 2012. Ecological assembly rules in plant communities: approaches, patterns and prospects. *Biological Reviews* **87**: 111–127.

Goulding, M. 1980. *The Fishes and the Forest: Explorations in Amazonian Natural History*. University of California Press, Berkeley, CA.

Grace, J. B. 2001. The roles of community biomass and species pools in the regulation of plant diversity. *Oikos* **92**: 193–207.

Grace, J. B., and D. Tilman (eds.). 1990. *Perspectives on Plant Competition*. Academic Press, San Diego, CA.

Graham, J. B. 1997. *Air Breathing Fishes*. Academic Press, San Diego, CA.

Greig-Smith, P. 1952. Use of random and contiguous quadrats in the study of the structure of plant communities. *Annals of Botany* 16: 293–316.

Greig-Smith, P. 1957. *Quantitative Plant Ecology*. Butterworths, London.

Grime, J. P. 1973a. Control of species density in herbaceous vegetation. *Journal of Environmental Management* 1: 151–167.

Grime, J. P. 1973b. Competitive exclusion in herbaceous vegetation. *Nature* 242: 344–347.

Grime, J. P. 1977. Evidence for the existence of three primary strategies in plants and its relevance to ecological and evolutionary theory. *The American Naturalist* 111: 1169–1194.

Grime, J. P. 1979. *Plant Strategies and Vegetation Processes*. Wiley, Chichester.

Grime, J. P. and R. Hunt. 1975. Relative growth rate: its range and adaptive influence in a local flora. *Journal of Ecology* 63: 393–422.

Grime, J. P., and S. Pierce. 2012. *The Evolutionary Strategies that Shape Ecosystems*. Wiley-Blackwell, Oxford.

Grime, J. P., G. Mason, A. V. Curtis, et al. 1981. A comparative study of germination characteristics in a local flora. *Journal of Ecology* 69: 1017–1059.

Grime, J. P., K. Thompson, R. Hunt, et al. 1997. Integrated screening validates primary axes of specialisation in plants. *Oikos* 79: 259–281.

Grubb, P. J. 1977. The maintenance of species-richness in plant communities: the importance of the regeneration niche. *Biological Reviews* 52: 107–145.

Grubb, P. J. 1986. Problems posed by sparse and patchily distributed species in species-rich plant communities. Pages 207–225 in *Community Ecology*. J. M. Diamond and T. J. Case (eds.). Harper and Row, New York.

Guo, Q., P. W. Rundel, and D. W. Goodall. 1998. Horizontal and vertical distribution of desert seed banks: patterns, causes, and implications. *Journal of Arid Environments* 38: 465–478.

Hacke, U. G., J. S. Sperry, W. T. Pockman, S. D. Davis and K. A. McCulloh. 2001. Trends in wood density and structure are linked to prevention of xylem implosion by negative pressure. *Oecologia* 126: 457–461.

Hall, C. A. S., J. A. Stanford and R. Hauer. 1992. The distribution and abundance of organisms as a consequence of energy balances along multiple environmental gradients. *Oikos* 65: 377–390.

Hamann, O. 1979. Regeneration of vegetation on Santa Fe and Pinta Islands, Galápagos, after the eradication of goats. *Biological Conservation* 15: 215–236.

Hamann, O. 1993. On vegetation recovery, goats and giant tortoises on Pinta Island, Galápagos, Ecuador. *Biodiversity and Conservation* **2**: 138–151.

Hansen, D. M., C. J. Donlan, C. J. Griffiths and K. J. Campbell. 2010. Ecological history and latent conservation potential: large and giant tortoises as a model for taxon substitutions. *Ecography* **33**: 272–284.

Harding, J. H. 2006. *Amphibians and Reptiles of the Great Lakes Region.* University of Michigan Press, Ann Arbor, MI.

Harper, J. L. 1977. *Population Biology of Plants.* Academic Press, London.

Harper, J. L., and I. H. McNaughton. 1962. The comparative biology of closely related species living in the same area. VII. Interference between individuals in pure and mixed populations of *Papaver* species. *New Phytologist* **61**: 175–188.

Harris, J. A., R. J. Hobbs, E. Higgs and J. Aronson. 2006. Ecological restoration and global climate change. *Restoration Ecology* **14**: 170–176.

Hartman, J. M. 1988. Recolonization of small disturbance patches in a New England salt marsh. *American Journal of Botany* **75**: 1625–1631.

He, F. 2010. Maximum entropy, logistic regression, and species abundance. *Oikos* **119**: 578–582.

Heck, K. L., and J. F. Valentine. 2007. The primacy of top-down effects in shallow benthic ecosystems. *Estuaries and Coasts* **30**: 371–381.

Hecnar, S. J., and R. T. M'Closkey. 1997. The effects of predatory fish on amphibian species richness and distribution. *Biological Conservation* **79**: 123–131.

Heffernan, M. 2011. *Willful Blindness: Why We Ignore the Obvious at Our Peril.* Doubleday Canada, Toronto.

Heinselman, M. L. 1973. Fire in the virgin forests of the Boundary Waters Canoe Area, Minnesota. *Quaternary Research* **3**: 329–382.

Henry, H. A. L., and Jeffries, R. L. 2009. Opportunist herbivores, migratory connectivity and catastrophic shifts in arctic coastal systems. Pages 85–102 in *Human Impacts on Salt Marshes: A Global Perspective.* B. R. Silliman, E. D. Grosholz and M. D. Bertness (eds.). University of California Press, Berkeley, CA.

HilleRisLambers, J., P. B. Adler, W. S. Harpole, J. M. Levine and M. M. Mayfield. 2012. Rethinking community assembly through the lens of coexistence theory. *Annual Review of Ecology, Evolution, and Systematics* **43**: 227–248.

Hoagland, B. W., and S. L. Collins. 1997. Gradient models, gradient analysis, and hierarchical structure in plant communities. *Oikos* **78**: 23–30.

Hubbell, S. P., and R. B. Foster. 1986. Biology, chance and history and the structure of the tropical rain forest tree communities. Pages 314–329 in *Community Ecology.* J. Diamond and T. J. Case (eds.). Harper and Row, New York.

Huston, M. A. 1979. A general hypothesis of species diversity. *The American Naturalist* **113**: 81–101.

Huston, M. A. 1994. *Biological Diversity: The Coexistence of Species on Changing Landscapes*. Cambridge University Press, Cambridge.

Hutchinson, G. E. 1958. Concluding remarks. *Cold Spring Harbor Symposium on Quantitative Biology* **22**: 415–427.

Hutchinson, G. E. 1959. Homage to Santa Rosalia or why are there so many kinds of animals? *The American Naturalist* **93**: 145–149.

Hutchinson, G. E. 1961. The paradox of the plankton. *The American Naturalist* **95**: 137–146.

Hutchinson, G. E. 1975. *A Treatise on Limnology. Volume 3. Limnological Botany*. Wiley, New York.

Jackson, J. B. C. 1981. Interspecific competition and species distributions: the ghosts of theories and data past. *American Zoologist* **21**: 889–901.

Jackson, J. B. C., and A. G. Coates. 1986. Life cycles and evolution of clonal (modular) animals. *Philosophical Transactions of the Royal Society of London, Series B* **313**: 7–22.

Jamil, T., W. A. Ozinga, M. Kleyer and C. J. F. ter Braak. 2013. Selecting traits that explain species–environment relationships: a generalized linear mixed model approach. *Journal of Vegetation Science* **24**: 988–1000.

Jaynes, E. T. 2003. *Probability Theory: The Logic of Science*. Cambridge University Press, Cambridge.

Jeffreys, H. 1973. *Scientific Inference*, 3rd ed. Cambridge University Press, Cambridge.

Jenny, H. 1941. *Factors of Soil Formation: A System of Quantitative Pedology*. McGraw-Hill, New York.

Jervis, R. A. 1969. Primary production in the freshwater marsh ecosystem of Troy Meadows, New Jersey. *Bulletin of the Torrey Botanical Club* **96**: 209–231.

Junk, W. J. 1984. Ecology of the várzea, floodplain of Amazonian white-water rivers. Pages 215–243 in *The Amazon Limnology and Landscape Ecology of a Mighty Tropical River and its Basin*. H. Sioli (ed.). Junk Publishers, The Hague.

Junk, W. J., M. G. M. Soares and U. Saint-Paul. 1997. The fish. Pages 385–408 in *The Central Amazon Floodplain: Ecology of a Pulsing System*. W. J. Junk (ed.). Springer-Verlag, Berlin.

Kattge, J., G. Bonisch, S. Diaz, et al. 2020. TRY plant trait database: enhanced coverage and open access. *Global Change Biology* **26**: 119–188.

Keddy, P. A. 1983a. Freshwater wetlands human-induced changes: indirect effects must also be considered. *Environmental Management* **4**: 299–302.

Keddy, P. A. 1983b. Shoreline vegetation in Axe Lake, Ontario: effects of exposure on zonation patterns. *Ecology* **64**: 331–344.

Keddy, P. A. 1985. Lakeshore plants in the Tusket River Valley, Nova Scotia: the distribution and status of some rare species including *Coreopsis rosea* and *Sabatia kennedyana. Rhodora* **87**: 309–320.

Keddy, P. A. 1987. Beyond reductionism and scholasticism in plant community ecology. *Vegetatio* **69**: 209211.

Keddy, P. A. 1990a. The use of functional as opposed to phylogenetic systematics: a first step in predictive community ecology. Pages 387–406 in *Biological Approaches and Evolutionary Trends in Plants*. S. Kawano (ed.). Academic Press, London.

Keddy, P. A. 1990b. Competitive hierarchies and centrifugal organization in plant communities. Pages 265–290 in *Perspectives on Plant Competition*. J. Grace and D. Tilman (eds.). Academic Press, New York.

Keddy, P. A. 1991a. Working with heterogeneity: an operator's guide to environmental gradients. Pages 181–201 *Ecological Heterogeneity* in J. Kolasa and S. T. A. Pickett (eds.). Springer Verlag, New York.

Keddy, P. A. 1991b. Reviewing a Festschrift: what are we doing with our scientific lives? Review of Diversity and Pattern in Scientific Communities (H. J. During et al. 1998. SPB Academic Publishing, The Hague). *The Journal of Vegetation Science* **2**: 419–424.

Keddy, P. A. 1992. Assembly and response rules: two goals for predictive community ecology. *Journal of Vegetation Science* **3**: 157–164.

Keddy, P. A. 1993. A pragmatic approach to functional ecology. *Functional Ecology* **6**: 621–626.

Keddy, P. A. 2001. *Competition*, 2nd ed. Kluwer, Dordrecht.

Keddy, P. A. 2005. Putting the plants back into plant ecology: six pragmatic models for understanding and conserving plant diversity. (Invited Review). *Annals of Botany* **95**: 1–13.

Keddy, P. A. 2009. Thinking big: a conservation vision for the southeastern coastal plain of North America. *Southeastern Naturalist* **8**: 213–226.

Keddy, P. A. 2010. *Wetland Ecology: Principles and Conservation*. Cambridge University Press, Cambridge.

Keddy, P. A. 2017. *Plant Ecology: Origins, Processes, Consequences*. Cambridge University Press, Cambridge.

Keddy, P. A. 2020. *Darwin Meets the Buddha: Human Nature, Buddha Nature, Wild Nature*. Sumeru Press, Ottawa.

Keddy, P. A., and B. Shipley. 1989. Competitive hierarchies in herbaceous plant communities. *Oikos* **54**: 234–241.

Keddy, P. A., L. H. Fraser and I. C. Wisheu. 1998. A comparative approach to examine competitive response of 48 wetland plant species. *Journal of Vegetation Science* **9**: 777–786.

Keddy, P. A., C. Gaudet and L. H. Fraser. 2000. Effects of low and high nutrients on the competitive hierarchy of 26 shoreline plants. *Journal of Ecology* **88**: 413–423.

Keddy, P. A., L. Smith, D. R. Campbell, M. Clark and G. Montz. 2006. Patterns of herbaceous plant diversity in southeastern Louisiana pine savannas. *Applied Vegetation Science* **9**: 17–26.

Keddy, P. A., D. Campbell, T. McFalls, et al. 2007. The wetlands of lakes Pontchartrain and Maurepas: past, present and future. *Environmental Reviews* **15**: 1–35.

Keddy, P. A., L. Gough, J. A. Nyman, et al. 2009. Alligator hunters, pelt traders, and runaway consumption of Gulf coast marshes: a trophic cascade perspective on coastal wetland losses. Pages 115–133 in *Human Impacts on Salt Marshes: A Global Perspective*. B. R. Silliman, E. D. Grosholz and M. D. Bertness (eds.). University of California Press, Berkeley, CA.

Kellman, M., and M. Kading. 1992. Facilitation of tree seedling establishment in a sand dune succession. *Journal of Vegetation Science* **3**: 679–688.

Kelly, R., K. Healy, M. Anand, et al. 2021. Climatic and evolutionary contexts are required to infer plant life history strategies from functional traits at a global scale. *Ecology Letters* **24**: 970–983.

Kempton, R. A. 1979. The structure of species abundance and the measurement of diversity. *Biometrics* **35**: 307–321.

Kenkel, N. C. 2006. On selecting an appropriate multivariate analysis. *Canadian Journal of Plant Science* **86**: 663–676.

Kenrick, P., and P. R. Crane. 1997. The origin and early evolution of plants on land. *Nature* **389**: 33–39.

Kershaw, K. A. 1973. *Quantitative and Dynamic Plant Ecology*, 2nd ed. Edward Arnold, London.

King, J. 1997. *Reaching for the Sun: How Plants Work*. Cambridge University Press, New York.

Kleyer, M., R. M. Bekker, I. C. Knevel, et al. 2008. The LEDA Traitbase: a database of life-history traits of the Northwest European flora. *Journal of Ecology* **96**: 1266–1274.

Kraft, N. J. B., R. Valencia and D. D. Ackerly. 2008. Functional traits and niche-based tree community assembly in an Amazonian forest. *Science* **322**: 580–582.

Kraft, N. J. B., L. S. Comita, J. M. Chase, et al. 2011. Disentangling the drivers of β diversity along latitudinal and elevational gradients. *Science* **333**: 1755–1758.

Kraft, N. J. B., P. B. Adler, O. Godoy, et al. 2015. Community assembly, coexistence and the environmental filtering metaphor. *Functional Ecology* **29**: 592–599.

Kramer, D. L., C. C. Lindsay, G. E. E. Moodie and E. D. Stevens. 1978. The fishes and the aquatic environment of the Central Amazon basin, with particular reference to respiratory patterns. *Canadian Journal of Zoology* **56**: 717–729.

Kramer-Walter, K. R., P. J. Bellingham, T. R. Millar, et al. 2016. Root traits are multidimensional: specific root length is independent from root tissue density and the plant economic spectrum. *Journal of Ecology* **104**: 1299–1310.

Kröpelin, S., D. Verschuren, A.-M. Lézine, et al. 2008. Climate-driven ecosystem succession in the Sahara: the past 6000 years. *Science* **329**: 765–768.

Lack, D. 1964. A long-term study of the Great Tit (*Parus major*). *Journal of Animal Ecology* **33**: 159–173.

Laliberté, E., B. Shipley, D. A. Norton and D. Scott. 2012. Which plant traits determine abundance under long-term shifts in soil resource availability and grazing intensity? *Journal of Ecology* **100**: 662–677.

Lane, N. 2010. *Life Ascending: Ten Great Inventions of Evolution*. W.W. Norton, New York.

Larcher, W. 1995. *Physiological Plant Ecology: Ecophysiology and Stress Physiology of Functional Groups*, 3rd ed. Springer-Verlag, New York.

Larcher, W. 2003. *Physiological Plant Ecology: Ecophysiology and Stress Physiology of Functional Groups*, 4th ed. Springer-Verlag, Berlin.

Larter, M., T. J. Brodribb, S. Pfautsch, et al. 2105. Extreme aridity pushes trees to their physical limits *Plant Physiology* **168**: 804–807.

Latham, P. J., L. G. Pearlstine and W. M. Kitchens. 1994. Species association changes across a gradient of freshwater, oligohaline, and mesohaline tidal marshes along the lower Savannah River. *Wetlands* **14**: 174–183.

Latham, R. E., J. Beyea, M. Benner, et al. 2005. *Managing White-tailed Deer in Forest Habitat from an Ecosystem Perspective: Pennsylvania Case Study*. Audubon Pennsylvania and Pennsylvania Habitat Alliance, Harrisburg, PA.

Laughlin, D. C. 2014. The intrinsic dimensionality of plant traits and its relevance to community assembly. *Journal of Ecology* **102**: 186–193.

Laughlin, D. C. 2018. Rugged fitness landscapes and Darwinian demons in trait-based ecology. *New Phytologist* **217**: 501–503.

Laughlin, D. C., and P. Z. Fulé. 2008. Wildland fire effects on understory plant communities in two fire-prone forests. *Canadian Journal of Forest Research* **38**: 133–142.

Laughlin, D. C., and D. E. Laughlin. 2013. Advances in modelling trait-based plant community assembly. *Trends in Plant Science* **18**: 584–593.

Laughlin, D. C., and C. Joshi. 2015. Theoretical consequences of trait-based environmental filtering for the breadth and shape of the niche: new testable hypotheses generated by the Traitspace model. *Ecological Modelling* **307**: 10–21.

Laughlin, D. C., and J. Messier. 2015. Fitness of multidimensional phenotypes in dynamic adaptive landscapes. *Trends in Ecology and Evolution* **80**: 487–496.

Laughlin, D. C., J. D. Bakker, M. T. Stoddard, et al. 2004. Toward reference conditions: wildfire effects on flora in an old-growth ponderosa pine forest. *Forest Ecology and Management* **199**: 137–152.

Laughlin, D. C., P. Z. Fulé, D. W. Huffman, J. Crouse and E. Laliberté. 2011. Climatic constraints on trait-based forest assembly. *Journal of Ecology* **99**: 1489–1499.

Laughlin, D. C., C. Joshi, P. M. van Bodegom, Z. A. Bastow and P. Z. Fulé. 2012. A predictive model of community assembly that incorporates intraspecific trait variation. *Ecology Letters* **15**: 1291–1299.

Laughlin, D. C., C. Joshi, S. J. Richardson, et al. 2015. Quantifying multimodal trait distributions improves trait-based predictions of species abundances and functional diversity. *Journal of Vegetation Science* **26**: 46–57.

Laughlin, D. C., R. T. Strahan, D. W. Huffman and A. J. Sánchez Meador. 2017. Using trait-based ecology to restore resilient ecosystems: historical conditions and the future of montane forests in western North America. *Restoration Ecology* **25**: S135–S146.

Laughlin, D. C., L. Chalmandrier, C. Joshi, et al. 2018a. Generating species assemblages for restoration and experimentation: A new method that can simultaneously converge on average trait values and maximize functional diversity. *Methods in Ecology and Evolution* **9**: 1764–1771.

Laughlin, D. C., R. T. Strahan, P. B. Adler and M. M. Moore. 2018b. Survival rates indicate that correlations between community-weighted mean traits and environments can be unreliable estimates of the adaptive value of traits. *Ecology Letters* **21**: 411–421.

Laughlin, D. C., J. R. Gremer, P. B. Adler, R. M. Mitchell and M. M. Moore. 2020. The net effect of functional traits on fitness. *Trends in Ecology and Evolution* **35**: 1037–1047.

Lavorel, S., K. Grigulis, S. McIntyre, et al. 2008. Assessing functional diversity in the field: methodology matters! *Functional Ecology* **22**: 134–147.

Leacock, S. 1911. *Nonsense Novels*. John Lane, London.

Lefebvre, J., S. W. Mockford and T. B. Herman. 2012. Ecology of a recently discovered population segment of Blanding's Turtles, *Emydoidea blandingii*, in Barren Meadow and Keddy Brooks, Nova Scotia. *The Canadian Field-Naturalist* **126**: 89–94.

Legendre, P., R. Galzin and M. L. Harmelin-Vivien. 1997. Relating behavior to habitat: solutions to the fourth-corner problem. *Ecology* **78**: 547–562.

Levin, H. L., and D. T. King. 2017. *The Earth Through Time*, 11th ed. Wiley, Hoboken, NJ.

Louisiana Natural Heritage Program. 2009. *The Natural Communities of Louisiana*. Louisiana Department of Wildlife and Fisheries, Baton Rouge, LA.

Loveless, C. M. 1959. A study of the vegetation in the Florida everglades. *Ecology* **40**: 1–9.

Lowe-McConnell, R. H. 1975. *Fish Communities in Tropical Freshwaters: Their Distribution, Ecology and Evolution*. Longman, London.

MacArthur, R. H. 1958. Population ecology of some warblers of northeastern coniferous forests. *Ecology* **39**: 599–619.

MacArthur, R. H. 1972. *Geographical Ecology: Patterns in the Distribution of Species*. Harper and Row, New York.

MacArthur, R. H., and E. O. Wilson. 1967. *The Theory of Island Biogeography*. Princeton University Press, Princeton, NJ.

Maglianesi, M. A., N. Blüthgen, K. Böhning-Gaese and M. Schleuning. 2014. Morphological traits determine specialization and resource use in plant–hummingbird networks in the neotropics. *Ecology* **95**: 3325–3334.

Magnuson, J. J., C. A. Paszkowski, F. J. Rahel and W. M. Tonn. 1989. Fish ecology in severe environments of small isolated lakes in northern Wisconsin. Pages 487–515 in *Freshwater Wetlands and Wildlife*. R. Sharitz and J. W. Gibbons (eds.). US Department of the Environment, Oak Ridge, TN.

Magnússon, B., S. H. Magnússon, E. Ólafsson and B. D. Sigurdsson. 2014. Plant colonization, succession and ecosystem development on Surtsey with reference to neighbouring islands. *Biogeosciences* **11**: 5521–5537.

Magurran, A. E., and B. J. McGill. 2011. *Biological Diversity: Frontiers in Measurement and Assessment*. Oxford University Press, Oxford.

Maherali, H., W. T. Pockman and R. B. Jackson. 2004. Adaptive variation in the vulnerability of woody plants to xylem cavitation. *Ecology* **85**: 2184–2199.

Major, J. 1951. A functional factorial approach to plant ecology. *Ecology* **32**: 392–412.

Marks, C. O., and H. C. Muller-Landau. 2007. Comment on "From plant traits to plant communities: a statistical mechanistic approach to biodiversity." *Science* **316**: 1425c.

Masese, F. O., N. Kitaka, J. Kipkemboi, et al. 2014. Macroinvertebrate functional feeding groups in Kenyan highland streams: evidence for a diverse shredder guild. *Freshwater Science* **33**: 435–450.

Mason, N. W. H., S. J. Richardson, D. A. Peltzer, et al. 2012. Changes in coexistence mechanisms along a long-term soil chronosequence revealed by functional trait diversity. *Journal of Ecology* **100**: 678–689.

Matheson, A. C., and C. A. Raymond. 1986. A review of provenance × environment interaction: its practical importance and use with particular reference to the tropics. *The Commonwealth Forestry Review* **65**: 283–302.

Matyas, C. 1994. Modeling climate change effects with provenance test data. *Tree Physiology* **14**: 797–804.

May, R. M. 1973. *Stability and Complexity in Model Ecosystems*. Princeton University Press, Princeton, NJ.

May, R. M. 1981. Patterns in multi-species communities. Pages 197–227 in *Theoretical Ecology*. R. M. May (ed.). Blackwell, Oxford.

May, R. M. 1986. The search for patterns in the balance of nature: advances and retreats. *Ecology* **67**: 1115–1126.

Mayfield, M. M., and J. M. Levine. 2010. Opposing effects of competitive exclusion on the phylogenetic structure of communities. *Ecology Letters* **13**: 1085–1093.

Mayfield, M. M., and D. B. Stouffer. 2017. Higher-order interactions capture unexplained complexity in diverse communities. *Nature Ecology and Evolution* **1**: 0062.

Mayr, E. 1982. *The Growth of Biological Thought: Diversity, Evolution, and Inheritance*. Belknap Press of Harvard University Press, Cambridge, MA.

McCarthy, M. A. 2007. *Bayesian Methods for Ecology*. Cambridge University Press, Cambridge.

McCune, B., and J. B. Grace. 2002. *Analysis of Ecological Communities*. MJM Software Design, Gleneden Beach, OR.

McFalls, T., P. A. Keddy, D. Campbell and G. Shaffer. 2010. Hurricanes, floods, levees, and nutria: vegetation responses to interacting disturbance and fertility regimes with implications for coastal wetland restoration. *Journal of Coastal Research* **26**: 901–911.

McGill, B. J., B. J. Enquist, E. Weiher and M. Westoby. 2006. Rebuilding community ecology from functional traits. *Trends in Ecology and Evolution* **21**: 178–185.

Meave, J., M. Kellman, A. MacDougall and J. Rosales. 1991. Riparian habitats as tropical forest refugia. *Global Ecology and Biogeography Letters* **1**: 69–76.

Médail, F., and P. Quézel. 1999. Biodiversity hotspots in the Mediterranean basin: setting global conservation priorities. *Conservation Biology* **13**: 1510–1513.

Merow, C., A. M. Latimer and J. A. Silander, Jr. 2011. Can entropy maximization use functional traits to explain species abundances? A comprehensive evaluation. *Ecology* **92**: 1523–1537.

Merow, C., J. P. Dahlgren, C. J. E. Metcalf, et al. 2014. Advancing population ecology with integral projection models: a practical guide. *Methods in Ecology and Evolution* **5**: 99–110.

Merritt, R. W., and K. W. Cummins (eds.). 1984. *An Introduction to the Aquatic Insects of North America*, 2nd ed. Kendall/Hunt Publishing, Dubuque, IA.

Miller, J. E. D., E. I. Damschen and A. R. Ives. 2018. Functional traits and community composition: a comparison among community-weighted means, weighted correlations, and multilevel models. *Methods in Ecology and Evolution* **10**: 415–425.

Mittelbach, G. G., and B. J. McGill. 2019. *Community Ecology*. Oxford University Press, Oxford.

Moles, A. T., D. D. Ackerly, C. O. Webb, et al. 2005. A brief history of seed size. *Science* **307**: 576–580.

Molina-Venegas, R., A. Aparicio, F. J. Pina, B. Valdés and J. Arroyo. 2013. Disentangling environmental correlates of vascular plant biodiversity in a Mediterranean hotspot. *Ecology and Evolution* **3**: 3879–3894 (plus corrigendum p. 4849).

Moolman, H. J., and R. M. Cowling. 1994. The impact of elephant and goat grazing on the endemic flora of South African succulent thicket. *Biological Conservation* **68**: 53–61.

Mooney, H. A., and E. L. Dunn. 1970. Convergent evolution of Mediterranean climate evergreen sclerophyll shrubs. *Evolution* **24**: 292–303.

Mora, C., D. P. Tittensor, S. Adl, A. G. B. Simpson and B. Worm. 2011. How many species are there on Earth and in the ocean? *PLoS Biology* **9**: e1001127.

Morton, J. K., and J. M. Venn. 1990. *A Checklist of the Flora of Ontario: Vascular Plants*. University of Waterloo, Waterloo, ON.

Mueller-Dombois, D., and H. Ellenberg. 1974. *Aims and Methods of Vegetation Ecology*. Wiley, New York.

Münkemüller, T., L. Gallien, L. J. Pollock, et al. 2020. Dos and don'ts when inferring assembly rules from diversity patterns. *Global Ecology and Biogeography* **29**: 1212–1229.

Muscarella, R., and M. Uriarte. 2016. Do community-weighted mean functional traits reflect optimal strategies? *Proceedings of the Royal Society B* **283**: 2015. 2434

Myers, J. A., and K. E. Harms. 2009. Seed arrival, ecological filters, and plant species richness: a meta-analysis. *Ecology Letters* **12**: 1250–1260.

Myers, J. A., and K. E. Harms. 2011. Seed arrival and ecological filters interact to assemble high-diversity plant communities. *Ecology* **92**: 676–686.

Myers, J. A., J. M. Chase, I. Jiménez, et al. 2013. Beta-diversity in temperate and tropical forests reflects dissimilar mechanisms of community assembly. *Ecology Letters* **16**: 151–157.

Myers, N., R. A. Mittermeier, C. G. Mittermeier, G. A. B. da Fonseca and J. Kent. 2000. Biodiversity hotspots for conservation priorities. *Nature* **403**: 853–858.

Myers, R. A., and Worm, B. 2003. Rapid worldwide depletion of predatory fish communities. *Nature* **423**: 280–283.

Niklas, K. J., B. H. Tiffney and A. H. Knoll. 1983. Patterns in vascular land plant diversification. *Nature* **303**: 614–616.

Niklas, K. J., B. H. Tiffney and A. H. Knoll. 1985. Patterns in vascular plant diversification: an analysis at the species level. Pages 97–128 in *Phanerozoic Diversity Pattern: Profiles in Macroevolution*. J. W. Valentine (ed.). Princeton University Press, Princeton, NJ.

Noss, R. F., and A. Y. Cooperrider. 1994. *Saving Nature's Legacy*. Island Press, Washington, DC.

Oksanen, L., S. D. Fretwell, J. Arruda and P. Niemelä. 1981. Exploitation ecosystems in gradients of primary productivity. *The American Naturalist* **118**: 240–261.

Oksanen, L., M. Aunapuu, T. Oksanen, et al. 1997. Outlines of food webs in a low arctic tundra landscape in relation to three theories on trophic dynamics. Pages 351–373 in *Multitrophic Interactions in Terrestrial Systems*. A. C. Gange and V.K. Brown (eds.). Blackwell Science, Oxford.

Oksanen, J., F. G. Blanchet, M. Friendly, et al. 2019. *Vegan: Community Ecology Package*. R package version 2.5-6. https://CRAN.R-project.org/package=vegan.

Oldham, M. J., W. D. Bakowsky and D. A. Sutherland. 1995. *Floristic Quality Assessment System for Southern Ontario*. Ministry of Natural Resources, Peterborough, ON.

Olson, D. M., E. Dinerstein, E. D. Wikramanayake, et al. 2001. Terrestrial ecoregions of the world: a new map of life on Earth. *Bioscience* **51**: 933–938.

O'Neil, T. 1949. *The Muskrat in the Louisiana Coastal Marshes*. Louisiana Department of Wildlife and Fisheries, New Orleans, LA.

Orloci, L. 1978. *Multivariate Analysis in Vegetation Research*, 2nd ed. Junk, The Hague.

Ott, J. P., J. Klimešová and D. C. Hartnett. 2019. The ecology and significance of below-ground bud banks in plants. *Annals of Botany* **123**: 1099–1118.

Ouellet, C. E., and Sherk, L. C. 1967. Woody ornamental plant zonation III: suitability map of the probable winter survival of ornamental trees and shrubs. *Canadian Journal of Plant Science* **47**: 351–358.

Paine, C. E. T., and K. E. Harms. 2009. Quantifying the effects of seed arrival and environmental conditions on tropical seedling community structure. *Oecologia* **160**: 139–150.

Pake, C. E., and D. L. Venable. 1996. Seed banks in desert annuals: implications for persistence and coexistence in variable environments. *Ecology* **77**: 1427–1435.

Parolin, P. 2009. Submerged in darkness: adaptations to prolonged submergence by woody species of the Amazonian floodplains. *Annals of Botany* **103**: 359–376.

Pärtel, M., M. Zobel, K. Zobel and E. van der Maarcl. 1996. The spccics pool and its relationship to species richness: evidence from Estonian plant communities. *Oikos* **75**: 111–117.

Pausas, J. G. 2015. Bark thickness and fire regime. *Functional Ecology* **29**: 315–327.

Pavord, A. 2005. *The Naming of Names: The Search for Order in the World of Plants*. Bloomsbury, New York.

Peet, R. K. 1974. The measurement of species diversity. *Annual Review of Ecology and Systematics* **5**: 285–307.

Peet, R. K., and D. J. Allard. 1993. Longleaf pine-dominated vegetation of the southern Atlantic and eastern Gulf Coast region, USA. Pages 45–81 in *Proceedings of the Tall Timbers Fire. Ecology Conference 18.* S.M Hermann (ed.). Tall Timbers Research Station, Tallahassee, FL.

Penfound, W. T., and E. S. Hathaway 1938. Plant communities in the marshlands of southeastern Louisiana. *Ecological Monographs* **8**: 1–56.

Pennings, S. C., T. H. Carefoot, E. L. Siska, M. E. Chase, and T. A. Page. 1998. Feeding preferences of a generalist salt-marsh crab: relative importance of multiple plant traits. *Ecology* **79**: 1968–1979.

Peres-Neto, P. R., S. Dray and C. J. F. ter Braak. 2017. Linking trait variation to the environment: critical issues with community-weighted mean correlation resolved by the fourth-corner approach. *Ecography* **40**: 806–816.

Pessanha, A. L. M., F. G. Araújo, R. E. M. C. C. Oliveira, A. Ferreira da Silva, and N. S. Sales. 2015. Ecomorphology and resource use by dominant species of tropical estuarine juvenile fishes. *Neotropical Icthyology* **13**: 401–412.

Peters, R. H. 1980. From natural history to ecology. *Perspectives in Biology and Medicine* **23**: 191–203.

Peters, R. H. 1992. *A Critique for Ecology*. Cambridge University Press, Cambridge.

Peterson, R. T. 1980. *A Field Guide to the Birds*, 4th ed. Houghton Mifflin, Boston, MA.

Pfeiffer, N. E. 1914. Morphology of *Thismia americana. Botanical Gazette* **57**: 122–135.

Pianka, E. R. 1981. Competition and niche theory. Pages 167–196 in *Theoretical Ecology*. R. M. May (ed.). Blackwell, Oxford.

Pianka, E. R. 1983. *Evolutionary Ecology*, 3rd ed. Harper and Row, New York.

Pickett, S. T. A., and P. S. White. 1985. *The Ecology of Natural Disturbance and Patch Dynamics*. Academic Press, Orlando, FL.

Pielou, E. C. 1975. *Ecological Diversity*. Wiley, New York.

Pielou, E. C., and R. D. Routledge. 1976. Salt marsh vegetation: latitudinal gradients in the zonation patterns. *Oecologia* **24**: 311–321.

Pimm, S. L. 2001. *The World According to Pimm: A Scientist Audits the Earth*. McGraw-Hill, New York.

Pistón, N., F. de Bello, A. T. C. Dias, et al. 2019. Multidimensional ecological analyses demonstrate how interactions between functional traits shape fitness and life history strategies. *Journal of Ecology* **107**: 2317–2328.

Pla, L., F. Casanoves and J. Di Rienzo. 2012. *Quantifying Functional Biodiversity*. Springer, Dordrecht.

Poelwijk, F. J., D. J. Kiviet, D. M. Weinreich and S. J. Tans. 2007. Empirical fitness landscapes reveal accessible evolutionary paths. *Nature* **445**: 383–386.

Pollock, L. J., W. K. Morris and P. A. Vesk. 2012. The role of functional traits in species distributions revealed through a hierarchical model. *Ecography* **35**: 716–725.

Power, M. E. 1992. Top-down and bottom-up forces in food webs: do plants have primacy? *Ecology* **73**: 733–746.

Preston, F. W. 1962a. The canonical distribution of commonness and rarity: Part I. *Ecology* **43**: 185–215.

Preston, F. W. 1962b. The canonical distribution of commonness and rarity: Part II. *Ecology* **43**: 410–432.

Purcell, A. S. T., W. G. Lee, A. J. Tanentzap and D. C. Laughlin. 2019. Fine root traits are correlated with flooding duration while aboveground traits are related to grazing in an ephemeral wetland. *Wetlands* **39**: 291–302.

Rabinowitz, D. 1981. Seven forms of rarity. Pages 205–217 in *The Biological Aspects of Rare Plant Conservation*. H. Synge (ed.). Wiley, Chichester.

Raunkiaer, C. 1908. The statistics of life-forms as a basis for biological plant geography. Translated from Danish and republished in 1934. Pages 111–147 in *The Life Forms of Plants and Statistical Plant Geography: Being the Collected Papers of Raunkiaer*. Clarendon Press, Oxford.

Raunkiaer, C. 1934. The *Life Forms of Plants and Statistical Plant Geography: Being the Collected Papers of Raunkiaer*. Translated from the Danish, French and German. Preface by A. G. Tansley. Clarendon Press, Oxford.

Raup, D. N. 1966. Geometric analysis of shell coiling: general problems. *Journal of Paleontology* **40**: 1178–1190.

Raven, P. H., R. F. Evert and S. E. Eichhorn. 2005. *Biology of Plants*, 7th ed. W. H. Freeman, New York.

Raymond, M. 1950. *Esquisse Phytogéographique du Quebec*. Jardin Botanique de Montréal, Montréal.

Reader, J. 1997. *Africa: A Biography of the Continent*. Hamish Hamilton, London.

Redford, K. H. 1992. The empty forest. *Bioscience* **42**: 412–422.

Renner, S. C., and W. Hoesel. 2017. Ecological and functional traits in 99 bird species over a large-scale gradient in Germany. *Data* **2**(12).

Reznicek, A. A., and P. M. Catling. 1989. Flora of Long Point, Regional Municipality of Haldimand-Norfolk, Ontario. *Michigan Botanist* **28**: 99–175.

Richards, P. W. 1996. *The Tropical Rain Forest. An Ecological Study*, 2nd edn. Cambridge University Press, Cambridge.

Rigler, F. H. 1982. Recognition of the possible: an advantage of empiricism in ecology. *Canadian Journal of Fisheries and Aquatic Sciences* **39**: 1323–1331.

Rigler, F. H., and R. H. Peters. 1995. *Science and Limnology*. Ecology Institute, Oldendorf/Lutie.

Ripple, W. J., J. A. Estes, R. L. Beschta, et al. 2014. Status and ecological effects of the world's largest carnivores. *Science* **343**: 1241484.

Rodkin, D. 1994. Searching for *Thismia*. *Chicago Reader* (September 22–28). www.chicagoreader.com/chicago/searching-for-thismia/Content?oid=885570.

Rodrigues, R. C., E. Hasul, J. C. Assis, et al. 2019. ATLANTIC BIRD TRAITS: a data set of bird morphological traits from the Atlantic forests of South America. *Ecology* **100**: e02647

Roland, A. E., and E. C. Smith. 1969. *The Flora of Nova Scotia*. Nova Scotia Museum, Halifax, NS.

Rosenzweig, M. L. 1995. *Species Diversity in Space and Time*. Cambridge University Press, Cambridge.

Roskov Y., L. Abucay, T. Orrell, et al. (eds.). 2018. *Species 2000 and ITIS Catalogue of Life, 2018 Annual Checklist*. www.catalogueoflife.org/annual-checklist/2018.

Roxburgh, S. H., and K. Mokany. 2007. Comment on "From plant traits to plant communities: a statistical mechanistic approach to biodiversity." *Science* **316**: 1425b.

Rüger, N., R. Condit, D. H. Dent, et al. 2020. Demographic trade-offs predict tropical forest dynamics. *Science* **368**: 165–168.

Saavedra, S., R. P. Rohr, J. Bascompte, et al. 2017. A structural approach for understanding multispecies coexistence. *Ecological Monographs* **87**: 470–486.

Salguero-Gómez, R., O. R. Jones, C. R. Archer, et al. 2015. The COMPADRE Plant Matrix Database: an open online repository for plant demography. *Journal of Ecology* **103**: 202–218.

Salisbury, E. J. 1942. *The Reproductive Capacity of Plants: Studies in Quantitative Biology*. G. Bell and Sons Ltd, London.

Savile, D. B. O. 1956. Known dispersal rates and migratory potentials as clues to the origin of the North American biota. *The American Midland Naturalist* **56**: 434–453.

Sculthorpe, C. D. 1967. *The Biology of Aquatic Vascular Plants*. Reprinted 1985. Edward Arnold, London.

Severinghaus, W. D. 1981. Guild theory development as a mechanism for assessing environmental impact. *Environmental Management* **5**: 187–190.

Shao, S, Q. Quan, T. Cai, et al. 2016. Evolution of body morphology and beak shape revealed by a morphometric analysis of 14 Paridae species. *Frontiers in Zoology* **13**: 30.

Shimwell, D. W. 1971. *The Description and Classification of Vegetation*. University of Washington Press, Seattle, WA.

Shipley, B. 2010. *From Plant Traits to Vegetation Structure: Chance and Selection in the Assembly of Ecological Communities*. Cambridge University Press, Cambridge.

Shipley, B., and P. A. Keddy. 1987. The individualistic and community-unit concepts as falsifiable hypotheses. *Vegetatio* **69**: 47–55.

Shipley, B., and Parent, M. 1991. Germination responses of 64 wetland species in relation to seed size, minimum time to reproduction and seedling relative growth rate. *Functional Ecology* **5**: 111–118.

Shipley, B., P. A. Keddy, D. R. J. Moore and K. Lemky. 1989. Regeneration and establishment strategies of emergent macrophytes. *Journal of Ecology* **77**: 1093–1110.

Shipley, B., D. Vile and É. Garnier. 2006. From plant traits to plant communities: a statistical mechanistic approach to biodiversity. *Science* **314**: 812–814.

Shipley, B., D. C. Laughlin, G. Sonnier and R. Otfinowski. 2011. A strong test of a maximum entropy model of trait-based community assembly. *Ecology* **92**: 507–517.

Shipley, B., C. E. T. Paine and C. Baraloto. 2012. Quantifying the importance of local niche-based and stochastic processes to tropical tree community assembly. *Ecology* **93**: 760–769.

Shurin, J. B. 2000. Dispersal limitation, invasion resistance, and the structure of pond zooplankton communities. *Ecology* **81**: 3074–3086.

Silliman, B. R., and Zieman, J. C. 2001. Top-down control of *Spartina alterniflora* production by periwinkle grazing in a Virginia salt marsh. *Ecology* **82**: 2830–2845.

Simberloff, D. 1984. The great god of competition. *The Sciences* **24**: 17–22.

Simpson, A. H., S. J. Richardson and D. C. Laughlin. 2016. Soil–climate interactions explain variation in foliar, stem, root and reproductive traits across temperate forests. *Global Ecology and Biogeography* **25**: 964–978.

Sinclair, A. R. E., C. J. Krebs, J. M. Fryxell, et al. 2000. Testing hypotheses of trophic level interactions: a boreal forest ecosystem. *Oikos* **89**: 313–328.

Siqueira-Souza, F. K., C. Bayer, W. H. Caldas, et al. 2017. Ecomorphological correlates of twenty dominant fish species of Amazonian floodplain lakes. *Brazilian Journal of Biology* **77**: 199–206.

Snodgrass, J. W., A. L. Bryan, Jr., and J. Burger. 2000. Development of expectations of larval amphibian assemblage structure in southeastern depression wetlands. *Ecological Applications* **10**: 1219–1229.

Sokal, R. R., and Sneath, P. H. A. 1963. *Principles of Numerical Taxonomy.* W. H. Freeman, New York.

Sonnier, G., B. Shipley and M.-L. Navas. 2010. Plant traits, species pools and the prediction of relative abundance in plant communities: a maximum entropy approach. *Journal of Vegetation Science* **21**: 318–331.

Sonnier, G., M.-L. Navas, A. Fayolle and B. Shipley. 2012. Quantifying trait selection driving community assembly: a test in herbaceous plant communities under contrasted land use regimes. *Oikos* **121**: 1103–1111.

Sörensen, L. 2007. *A Spatial Analysis Approach to the Global Delineation of Dryland Areas of Relevance to the CBD Programme of Work on Dry and Subhumid Lands.* UN Environment Programme World Conservation Monitoring Centre (UNEP-WCMC), Cambridge.

Spasojevic, M. J., and K. N. Suding. 2012. Inferring community assembly mechanisms from functional diversity patterns: the importance of multiple assembly processes. *Journal of Ecology* **100**: 652–661.

Stearns, S. C. 1992. *The Evolution of Life Histories.* Oxford University Press, Oxford.

Stevens, P. W., S. L. Fox and C. L. Montague. 2006. The interplay between mangroves and saltmarshes at the transition between temperate and subtropical climate in Florida. *Wetlands Ecology and Management* **14**: 435–444.

Strimbeck, G. R., P. G. Schaberg, C. G. Fossdal, W. P. Schröder and T. D. Kjellsen. 2015. Extreme low temperature tolerance in woody plants. *Frontiers in Plant Science* **6**: 884–898.

Stuart, S. A., B. Choat, K. C. Martin, N. M. Holbrook and M. C. Ball. 2007. The role of freezing in setting the latitudinal limits of mangrove forests. *New Phytologist* **173**: 576–583.

Svensson, E. 2016. Adaptive landscapes. in *Oxford Bibliographies Online: Evolutionary Biology*. D. J. Futuyma (ed.). Oxford University Press, New York.

Swenson, N. G. 2019. *Phylogenetic Ecology: A History, Critique and Remodeling*. University of Chicago Press, Chicago, IL.

Takhtajan, A. 1986. *Floristic Regions of the World*. Translated by T. J. Crovello. University of California Press, Berkeley, CA.

Tanentzap A. J., and W. G. Lee. 2017. Evolutionary conservatism explains increasing relatedness of plant communities along a flooding gradient. *New Phytologist* **213**: 634–644.

Tanentzap A. J., W. G. Lee, A. Monks, et al. 2014. Identifying pathways for managing multiple disturbances to limit plant invasions. *Journal of Applied Ecology* **51**: 1015–1023.

Tansley, A. G. 1914. Presidential Address [to the British Ecological Society]. *Journal of Ecology* **2**: 194–203.

Temperton, V. M., R. J. Hobbs, T. Nuttle and S. Halle (eds.). 2004. *Assembly Rules and Restoration Ecology: Bridging the Gap Between Theory and Practice*. Island Press, Washington, DC.

ter Braak, C. J., P. Peres-Neto and S. Dray. 2017. A critical issue in model-based inference for studying trait-based community assembly and a solution. *PeerJ* **5**: e2885.

The Plant List. 2013. Version 1.1. A working list of all plant species. www.theplant list.org.

Thirgood, J. V. 1981. *Man and the Mediterranean Forest: A History of Resource Depletion*. Academic Press, London.

Thompson, D. Q., R. L. Stuckey and E. B. Thompson. 1987. *Spread, Impact, and Control of Purple Loosestrife* (Lythrum salicaria) *in North American Wetlands*. US Department of the Interior, Fish and Wildlife Service, Washington, DC.

Thompson, K., and M. A. McCarthy. 2008. Traits of British alien and native urban plants. *Journal of Ecology* **96**: 853–859.

Thompson, K., S. R. Band and J. G. Hodgson. 1993. Seed size and shape predict persistence in soil. *Functional Ecology* **7**: 236–241.

Thuiller, W., S. Lavorel, G. Midgley, S. Lavergne and T. Rebelo. 2004. Relating plant traits and species distributions along bioclimatic gradients for 88 Leucadendron taxa. *Ecology* **85**: 1688–1699.

Tomlinson, P. B. 1986. *The Botany of Mangroves*. Cambridge University Press, Cambridge.

Tonn, W. M., and J. J. Magnuson. 1982. Patterns in the species composition and richness of fish assemblages in northern Wisconsin lakes. *Ecology* **63**: 1149–1166.

Toti, D. S., F. A. Coyle and J. A. Miller. 2000. A structured inventory of Appalachian grass bald and heath bald spider assemblages and a test of species richness estimator performance. *Journal of Arachnology* **28**: 329–345.

Toussaint, A., N. Charpin, S. Brosse and S. Villéger. 2016. Global functional diversity of freshwater fish is concentrated in the Neotropics while functional vulnerability is widespread. *Scientific Reports* **6**: 22125.

Tozer, R. 2012. *Birds of Algonquin Park*. The Friends of Algonquin Park, Whitney, ON.

Treseder, K. K., and J. T. Lennon. 2015. Fungal traits that drive ecosystem dynamics on land. *Microbiology and Molecular Biology Reviews* **79**: 243–262.

Tuchman, B. W. 1984. *The March of Folly: From Troy to Vietnam*. Knopf, New York.

Turner, R. E., and N. N. Rabelais. 2003. Linking landscape and water quality in the Mississippi River Basin for 200 years. *BioScience* **53**: 563–572.

Turner, R. M., S. M. Alcorn, G. Oli and J. A. Booth. 1966. The influence of shade, soil, and water on saguaro seedling establishment. *Botanical Gazette* **127**: 95–102.

Tursch, B. 1997. Spiral growth: the "museum of all shells" revisited. *Journal of Molluscan Studies* **63**: 547–554.

Vallentyne, J. R. 1974. *The Algal Bowl: Lakes and Man*. Department of the Environment, Fisheries and Marine Service, Ottawa.

van der Heijden, M. G. A., and T. R. Horton. 2009. Socialism in soil? The importance of mycorrhizal fungal networks for facilitation in natural ecosystems. *Journal of Ecology* **97**: 1139–1150.

van der Pijl, L. 1982. *Principles of Dispersal in Higher Plants*, 3rd ed. Springer-Verlag, Berlin.

van der Valk, A. G. 1981. Succession in wetlands: a Gleasonian approach. *Ecology* **62**: 688–696.

van der Valk, A. G. 1989. *Northern Prairie Wetlands*. Iowa State University Press, Ames, IA.

Visser, J. M., S. D. Steyer, G. P. Shaffer, et al. 2004. Habitat switching module. Pages C–143–159 in *US Army Corps of Engineers and State of Louisiana Ecosystem Restoration Study Louisiana Coastal Area (LCA), Louisiana*. US Army Corps of Engineers, New Orleans, LA.

Vredenburg, V. T. 2004. Reversing introduced species effects: experimental removal of introduced fish leads to rapid recovery of a declining frog. *Proceedings of the National Academy of Science* **101**: 7646–7650.

Wallace, A. R. 1876. *The Geographical Distribution of Animals. With a Study of the Relations of Living and Extinct Faunas as Elucidating the Past Changes of the Earth's Surface*. Harper and Brothers, New York.

Wardle, D. A. 2002. *Communities and Ecosystems: Linking the Aboveground and Belowground Components*. Princeton University Press, Princeton, NJ.

Warton, D. I., B. Shipley and T. Hastie. 2015. CATS regression: a model-based approach to studying trait-based community assembly. *Methods in Ecology and Evolution* **6**: 389–398.

Webb, C. T., J. A. Hoeting, G. M. Ames, M. I. Pyne and N. LeRoy Poff. 2010. A structured and dynamic framework to advance traits-based theory and prediction in ecology. *Ecology Letters* **13**: 267–283.

Weiher, E., and P. A. Keddy. 1995. Assembly rules, null models, and trait dispersion: new questions from old patterns. *Oikos* **74**: 159–164.

Weiher, E., and P. A. Keddy (eds.). 1999a. *Ecological Assembly Rules: Perspectives, Advances, Retreats*. Cambridge University Press, Cambridge.

Weiher, E., and P. A. Keddy. 1999b. Assembly rules as general constraints on community composition. In *Ecological Assembly Rules: Perspectives, Advances, Retreats*. E. Weiher and P. Keddy (eds.). Cambridge University Press, Cambridge.

Weiher, E., A. van der Werf, K. Thompson, et al. 1999. Challenging Theophrastus: a common core list of plant traits for functional ecology. *Journal of Vegetation Science* **10**: 609–620.

Weldon, C. W., and W. L. Slauson. 1986. The intensity of competition versus its importance: an overlooked distinction and some implications. *Quarterly Review of Biology* **61**: 23–44.

Weller, M. W. 1999. *Wetland Birds: Habitat Resources and Conservation Implications*. Cambridge University Press, Cambridge.

Welty, J. C. 1982. *The Life of Birds*, 3rd ed. Saunders College Publishing, New York.

Westoby, M., D. S. Falster, A. T. Moles, P. A. Vesk and I. J. Wright. 2002. Plant ecological strategies: some leading dimensions of variation between species. *Annual Review of Ecology and Systematics* **33**: 125–159.

Wheelwright, N. T., and J. D. Rising. (2008). Savannah Sparrow (*Passerculus sandwichensis*), version 2.0. In *The Birds of North America*. P. G. Rodewald, (ed.). Cornell Lab of Ornithology, Ithaca, NY.

White, D. J. 2016. *Plants of Lanark County, Ontario – 2016 Edition*. www.lanark flora.com/.

White, P. S., S. P. Wilds and G. A. Thunhorst. 1998. Southeast. Pages 255–314 in *Status and Trends of the Nation's Biological Resources* (2 Vols.). M. J. Mac, P. A. Opler, C. E. Puckett Haecker and P. D Doran (eds.). U.S. Department of the Interior, U.S. Geological Survey, Reston, VA.

Whitmore, T., R. Peralta and K. Brown. 1985. Total species count in a Costa Rican tropical rain forest. *Journal of Tropical Ecology* 1: 375–378.

Whittaker, R. H. 1965. Dominance and diversity in land plant communities. *Science* 147: 250–260.

Wiens, J. A. 1983. Avian community ecology: an iconoclastic view. Pages 355–403 in *Perspectives in Ornithology, Essays Presented for the Centennial of the American Ornithologists' Union*. A. H. Brush and G. A. Clark, Jr. (eds.). Cambridge University Press, Cambridge.

Wilbur H. M. 1984. Complex life cycles and community organization in amphibians. Pages 195–225 in *A New Ecology: Novel Approaches to Interactive Systems*. P. W. Price, C. N. Slobodchikoff, and W. S. Gaud (eds.). Wiley, New York.

Williams, A. P., E. R. Cook, J. E. Smerdon, et al. 2020. Large contribution from anthropogenic warming to an emerging North American megadrought. *Science* 368: 314–318.

Williams, G. C. 1975. *Sex and Evolution*. Princeton University Press, Princeton, NJ.

Williams, M. 1989. The lumberman's assault on the southern forest, 1880–1920. Pages 238–288 in *Americans and Their Forests: A Historical Geography*. Williams, M. (ed.). Cambridge University Press, Cambridge.

Wisheu, I. C., and P. A. Keddy. 1989. The conservation and management of a threatened coastal plain plant community in eastern North America (Nova Scotia, Canada). *Biological Conservation* 48: 229–238.

Wisheu, I. C., and P. A. Keddy. 1992. Competition and centrifugal organization of ecological communities: theory and tests. *Journal of Vegetation Science* 3: 147–156.

Wisheu, I. C., and P. A. Keddy. 1994. The low competitive ability of Canada's Atlantic coastal plain shoreline flora: implications for conservation. *Biological Conservation* 68: 247–252.

Wisheu, I. C., C. J. Keddy, P. A. Keddy and N. M. Hill. 1994. Disjunct Atlantic coastal plain species in Nova Scotia: distribution, habitat and conservation priorities. *Biological Conservation* **68**: 217–224.

Witman, J. D., R. J. Etter and F. Smith. 2004. The relationship between regional and local species diversity in marine benthic communities: a global perspective. *Proceedings of the National Academic of Sciences* **101**: 15664–15669.

Wong, M. K. L., B. Guenard and O. T. Lewis. 2018. Trait-based ecology of terrestrial arthropods. *Biological Reviews* **94**: 992–1022.

Wood, K. A., M. T. O'Hare, C. McDonald, et al. 2017. Herbivore regulation of plant abundance in aquatic ecosystems. *Biological Reviews* **92**: 1128–1141.

Woodward, F. I. 1987. *Climate and Plant Distribution*. Cambridge University Press, Cambridge.

Wootton, R. J. 1990. Biotic interaction. II. Competition and mutualism. Pages 216–237 in *Ecology of Teleost Fishes*. R. J. Wootton (ed). Chapman and Hall, London.

World Wildlife Fund for Nature. 2014. *Mysterious Mekong: New Species Discoveries 2012–2013*. WWF-Greater Mekong, Bangkok.

Worthy, S. J., D. C. Laughlin, J. Zambrano, et al. 2020. Alternative designs and tropical tree seedling growth performance landscapes. *Ecology* **101**: e03007.

Wright, D. K. 2017. Humans as agents in the termination of the African humid period. *Frontiers in Earth Science* **5**.

Wright, I. J., P. B. Reich, M. Westoby, et al. 2004. The worldwide leaf economics spectrum. *Nature* **428**: 821–827.

Wright, J. P., and J. D. Fridley. 2010. Biogeographic synthesis of secondary succession rates in eastern North America. *Journal of Biogeography* **37**: 1584–1596.

Wright, S. 1932. The roles of mutation, inbreeding, crossbreeding and selection in evolution. Pages 356–366 in *Proceedings of the Sixth International Congress of Genetics. Vol. 1, Transactions and General Addresses*. D. F. Jones (ed). Genetics Society of America, Austin, TX.

Yodzis, P. 1978. *Competition for Space and the Structure of Ecological Communities*. Springer-Verlag, Berlin.

Yodzis, P. 1986. Competition, mortality, and community structure. Pages 480–492 in *Community Ecology*. J. Diamond and T. J. Case (eds.). Harper and Row, New York.

Zobel, M. 2016. The species pool concept as a framework for studying patterns of plant ecology. *Journal of Vegetation Science* **27**: 8–18.

Index

Page numbers in italics refer to figures, page numbers in bold to tables.

For EU product safety concerns, contact us at Calle de José Abascal, 56–1°,
28003 Madrid, Spain or eugpsr@cambridge.org.

www.ingramcontent.com/pod-product-compliance
Ingram Content Group UK Ltd.
Pitfield, Milton Keynes, MK11 3LW, UK
UKHW040949090126
466816UK00019B/330